Recent Oil Spill Challenges That Require More Attention

Edited by Maged Marghany

Published in London, United Kingdom

Recent Oil Spill Challenges That Require More Attention
http://dx.doi.org/10.5772/intechopen.102299
Edited by Maged Marghany

Contributors
Jerry Danwazan, Adesina Mutiu Adeleye, Trust-Paul Bakyu, Mamidak M. Iliya, Joseph Kehinde Egunjobi, Singanamalla Vijayakumar, Len J. Pietrafesa, Paul T. Gayes, Shaowu Bao, Farid Askari, Veronika Veľková, Helena Hybská, Tatiana Bubeníková, Maged Marghany

First published in London, United Kingdom, 2023 by IntechOpen
IntechOpen is the global imprint of INTECHOPEN LIMITED, registered in England and Wales, registration number: 11086078, 5 Princes Gate Court, London, SW7 2QJ, United Kingdom

British Library Cataloguing-in-Publication Data
A catalogue record for this book is available from the British Library

Additional hard and PDF copies can be obtained from orders@intechopen.com

Recent Oil Spill Challenges That Require More Attention
Edited by Maged Marghany
p. cm.
Print ISBN 978-1-83969-114-0
Online ISBN 978-1-83969-115-7
eBook (PDF) ISBN 978-1-83969-117-1

Meet the editor

Distinguished Prof. Dr. Maged Marghany, the innovator of a novel theory named "Quantized Marghany's Front," obtained a BSc in Physical Oceanography from the University of Alexandria, Egypt, an MSc in Physical Oceanography from the University Pertanian Malaysia, a Ph.D. in Environmental Remote Sensing from the Universiti Putra Malaysia, and a post-doctorate degree in Radar Remote Sensing from the International Institute for Aerospace Survey and Earth Sciences, the Netherlands. The prestigious Universidade Estadual de Feira de Santana, Universidade Federal da Bahia, and Universidade Federal de Pernambuco, Brazil, ranked him as the first global scientist in the field of oil spill detection and mapping during the last 50 years. Dr. Marghany is currently a director of Global Geoinformation Sdn.Bhd. His research specializes in microwave remote sensing and remote sensing for mineralogy detection and mapping. Previously, he was a deputy director in research and development at the Institute of Geospatial Science and Technology and the Department of Remote Sensing, both at Universiti Teknologi Malaysia. He has published more than 250 papers in international conferences and journals and 8 books. He is active in the *International Journal of Geoinformatics* and the International Society for Photogrammetry and Remote Sensing (ISPRS). Dr. Marghany was listed among the world's top 2% of scientists by Stanford University, USA, in 2020, 2021, and 2022. Two of his books were listed seventh and fortieth among the best genetic algorithm books of all time.

Contents

Preface

The tragedy of the 2010 Gulf of Mexico oil spill won't soon be forgotten. It serves as a reminder of how much more needs to be done in terms of protecting the environment from disasters. This book provides a comprehensive overview of oil spills from various perspectives. These perspectives shed light on the potential environmental effects of oil spills, how to identify them from space, and how to mitigate or prevent them.

Chapter 1 examines the use of key remote sensing to track oil spills and discusses the greatest obstacles to remotely monitoring oil spills.

Chapter 2 discusses the relationship between thermal maturity evolution and hydrocarbon cracking. The former is a process whereby the chemical composition of organic matter changes as it is subjected to increasing levels of heat over time. The latter is a chemical reaction that occurs when organic molecules are exposed to high temperatures, which breaks them down into smaller components. These two processes are related in that thermal maturity evolution is a prerequisite for hydrocarbon cracking to occur. The relationship between thermal maturity evolution and hydrocarbon cracking is critical in petroleum geology, as it can give insight into the formation of oil reserves. By understanding the relationship between thermal maturity evolution and hydrocarbon cracking, petroleum geologists can better assess potential areas for oil exploration. The Anambra Basin in Nigeria's Lower Benue Trough serves as the case study in this chapter. According to the chapter, post-maturity status could be associated with deeper burial depths, possibly as a result of the effect of Santonian tectonic episodes.

Chapter 3 discusses mechanical techniques for cleaning up oil spills. It describes the use of sorbents, oil skimmers, and containment booms for accidental oil spills. It compares the effectiveness of various sorbent types using experiments and studies. Non-polar hydrocarbons were found in purified water, which demonstrated that additional cleaning techniques were needed because the water did not meet the standards for surface water. The chapter also tests an absorption sock's sorption capacity, which is beneficial for smaller oil spills because it combines the containment of oil and the recovery of oil compounds.

Chapter 4 provides examples of local and nonlocal oil spills and discusses how they affect South Carolina's coastal waters and the continental margins of the United States. In 2010, there was a sizeable oil spill in the Gulf of Mexico, which has a huge number of oil drilling platforms and daily oil tanker traffic. The United States Administration lifted its ban on oil drilling in the coastal waters of North Carolina in 2017. Thus, two inquiries are raised. First, could the coast of Carolina be affected by an oil spill in the Gulf? Second, under typical environmental conditions, could an oil spill reach the beaches if it occurred off the coast of North Carolina in the future? This chapter offers answers to both of these questions.

Finally, Chapter 5 discusses detecting and identifying oil spills in coastal regions or marine surroundings. Normally, oil spills in coastal regions impact the characteristics of environmental activities, which are monitored through several radar satellites and sensors. Thus, researchers are developing several of these approaches for detecting and identifying spills in these areas.

The chapter also discusses the effects of contemporary environmental bio-systems, their control measurement approaches, and their surveillance operations. These operations include characterizations of oil spills and the quality of their impacts. Eventually, the synthetic aperture radar (SAR) image region classification based on its feature extraction will be utilized to detect oil spills. The dark region selection in the image can be used to monitor this using remote sensing techniques.

I wish to convey my appreciation to the staff at IntechOpen, especially Author Service Manager Ms. Dolores Kuzelj, for their assistance throughout the publication process.

Dr. Maged Marghany
Distinguished Professor,
Global Geoinformation, Sdn. Bhd.
Kuala Lumpur, Malaysia

Chapter 1

Introductory Chapter: Issues with Oil Spills and Remote Monitoring

Maged Marghany

1. Introduction

The oil spill is a popular topic for pollution discussion because of the extent of the environmental harm it can produce. Additionally, it may have a variety of negative effects on ecosystems and economies, ranging from immediate extinction of species and habitat destruction to long-term water contamination effects. These outcomes can be disastrous and can include everything from long-term water pollution to irreversible changes in species populations and habitat destruction. This introductory chapter would be demonstrated a variety of other concepts in addition to oil spills, in the author's opinion.

1.1 What is meant by an oil spill?

What exactly is oil? In line with the basic definition, oil is an organic compound that is soluble or readily soluble in water but not readily soluble in other liquids and is found in crude oil. An oil spill occurs when oil floats on the surface of bodies of water and is carried by the wind, currents, and tides [1–5].

The most visible source of oil pollution in the marine environment is operational oil discharges and spills from ships, particularly tankers, offshore platforms, and pipelines. Take the Amoco Cadiz oil tanker as an example (**Figure 1**). In 1978, this tanker ran aground off the French coast, causing 68.7 million gallons of oil to spill.

Figure 1.
Sinking oil tanker Amoco Cadiz.

IntechOpen

Large spills like these are unusual occurrences. Additionally, an oil transfer accident caused the oil tanker Mega-Borg (**Figure 2**) to spill 5.1 million gallons of oil [1, 6].

Ixtox 1 exploration well blowout in 1979. In 1980, when workers were able to stop the blowout, 140 million gallons of oil were estimated to have leaked into the ocean (**Figure 3**). Only the intentional oil spills that put an end to the Kuwait-Iraq war in 1991 are smaller than this, making it the second-largest spill in history (**Figure 4**).

Figure 2.
Explosion of tanker mega-Borg.

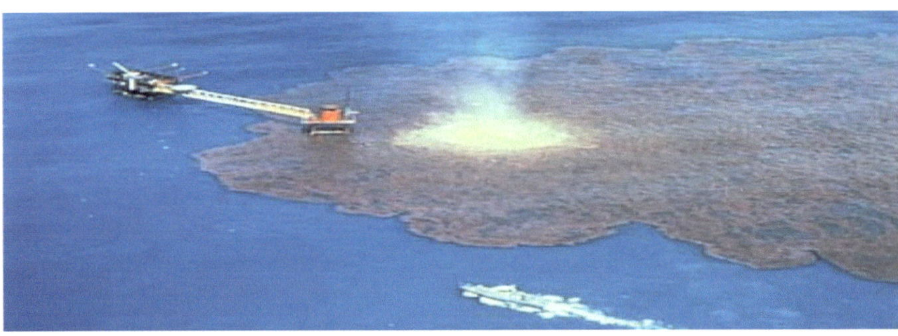

Figure 3.
Exploratory well Ixtox 1 in 1979.

Figure 4.
Exploratory of oil wells during the Kuwait-Iraq war of 1991.

In fact, during times of conflict, one nation may choose to pour tons of oil into the oceans of the opposing nation [1, 4, 6, 7].

2. Oil spill behavior in marine environments

The environmental impact of an oil spill will depend on how quickly it spreads. The majority of oils have the propensity to spread horizontally, leaving a "slick"—a smooth and slick surface—on top of the water. Surface tension, specific gravity, and viscosity are factors that influence how easily an oil spill can spread. The degree of attraction between a liquid's surface molecules is measured by surface tension. The likelihood that an oil spill will continue increases with the surface tension of the oil [6–10]. Even without the aid of wind and water currents, the oil will spread if its surface tension is low. Oil is more likely to spread in warmer waters than in extremely cold ones since increased temperatures can reduce a liquid's surface tension [1–8, 11–15].

The density of a substance as compared to the density of water is known as its specific gravity. Most oils float on top of the water because they are lighter than water. However, if the lighter components of the oil evaporate, the specific gravity of an oil spill may rise. Animal fats, vegetable oils, and heavier oils may interact with sediments or rocks at the bottom of a body of water, sink, and form tar balls. The measure of a liquid's flow resistance is its viscosity. The tendency of the oil to settle in one place increases with its viscosity [1, 6, 8, 16].

3. Problem of technical methods for oil spill detection

Recent advancements in remote sensing technology have made them an important tool for surveying and detecting marine pollution, which helps to better detect oil spills. Ships, aircraft, and satellites are some of the many tools available to detect and monitor oil spills [17, 18]. When ships are outfitted with navigation radars, they can detect oil spills at sea, say in restricted areas of 2500 m × 2500 m (**Figure 5**). As an alternative, the main methods for observing oil pollution in the ocean are aircraft and satellites [1, 3, 19–21].

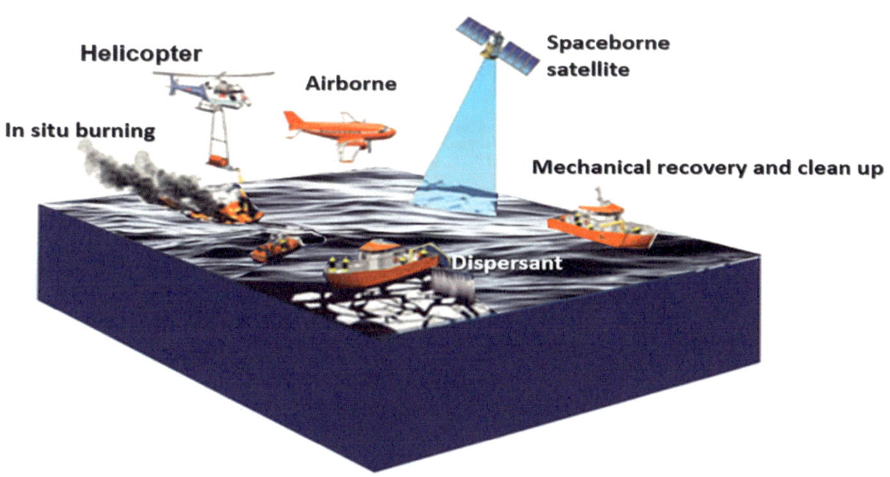

Figure 5.
Different tools for oil spill detectionz.

4. Monitoring an oil spill with optical remote sensing: potential

Research on the utilization of remote sensing technology for oil spill pollution detection has been ongoing for more than a decade. Optical and microwave sensors have been examined in several studies for oil spill detection and monitoring in several coastal water. Because of its wide area coverage and day and night all-weather capabilities, SAR deployed on satellites is now an essential instrument in oil spill monitoring. However, the majority of research on airborne remote sensing methods is not included. Although there have been a few attempts to use optical data, the main satellite or airborne data source primarily used for oil spill detection is the microwave sensor [1, 6, 22].

The most popular technique for remote sensing is optical. Due to their low cost and availability as commercial-off-the-shelf (COTS) items, cameras—both still and video—are widely used. Now, affordable digital single-reflex (SLR) cameras and camcorders are readily available. In the visible range (between 400 and 700 nm), oil has a higher surface reflectance than water, but it exhibits fewer general absorption tendencies [6, 22].

The oil that is "optically thick" absorbs solar energy and releases it again as thermal energy, primarily in the 8–12 μm spectral range [18–20]. Infrared images show that thick oil is hot, intermediate thicknesses are cool, and thin oil or sheen cannot be seen. In the evening, the opposite is seen. Although the precise thicknesses at which these transitions take place are unknown, scientific evidence indicates that the minimum detectable layer occurs between 10 and 70 μm and that the transition between the hot and cool layers happens between 50 and 150 μm. Due to the difference in emissivity between oil (0.94–0.97) and water, when the oil and water are at the same actual temperature, the oil will appear to be cooler (0.98). The infrared temperature difference between oil and water can be used to detect oil spills, and the magnitude of the difference is correlated with the thickness of the layer [1, 17, 18, 22].

Therefore, infrared sensors in the 8–12 μm range are now far more reliable and accurate than traditional infrared scanners. In this view, the "thermal infrared region," with wavelengths between 8 and 12 μm, is where infrared remote sensing has most frequently been used. Mid-band IR system tests (3–5 μm) have shown that these sensors might be useful. No spectral structure is present in this region, according to specific studies in the thermal infrared (812 μm) [17, 18, 22].

However, oil detection using infrared technology is not foolproof due to potential interferences and false positives, so the use of both IR and UV together can offer a more conclusive sign of the presence of oil than either method by itself. UV sensors are not used in operational response modes and would not play a significant role unless they are used in conjunction with IR technologies [1, 6, 17, 18, 22]. The Moderate-Resolution Imaging Spectroradiometer (MODIS) instrument may be used to keep track of oil spills. MODIS has a broad spectral range and two bands with a moderate resolution of 250 m and 500 m. On the other hand, using multiple wavelengths can give you more information to tell the difference between oil spills and slicks caused by algal blooms. However, in a tropical region like Malacca Straits, where there is a lot of cloud cover, the MODIS data has a lot of problems because of heavy cloud cover [17, 18, 20, 21]. The use of hyperspectral sensors for oil spill monitoring has the potential to offer precise material identification and a more accurate assessment of their abundance. A hyperspectral sensor's more than 200 wavelengths allow for the exploitation of the spectral signature of oil and the differentiation of various oil types. This can reduce the frequency of false alarms caused by ocean features resembling oil in appearance and color [1, 22].

In this regard, a signature matching method based on airborne hyperspectral imaging is more precise than conventional techniques, where analysis is based on a visual interpretation of the oil's color and its appearance in the satellite image. There is not a commercial hyperspectral sensor in orbit right now. One example of a space-borne technology demonstrator that was launched in 2000 is the NASA EO-1 Hyperion hyperspectral sensor. However, its narrow swath width of only 7.5–100 km is its main disadvantage [20, 22].

The oil absorbs solar energy and releases some of it as thermal energy back into the atmosphere. This oil cannot be detected by IR sensors, which perceive thick oil slicks as hot and intermediate thicknesses of oil as cool. A thick spill can appear cooler than the surrounding water at night because it dissipates heat more quickly. Oil can quickly absorb and release thermal energy, whereas water has a slower rate of heat absorption and release. Oil slicks can cool off more quickly than the surrounding water as a result, and IR sensors can detect this temperature difference. To identify the presence of oil slicks in water bodies, IR sensors are crucial tools [22].

The NOAA Advanced Very High-Resolution Radiometer (AVHRR) has visible and infrared sensors with early detection and monitoring capabilities for oil spills. The 1991 Persian Gulf War's oil spills were investigated. The oil spills might not have a temperature signature that is noticeably different from the surrounding water at night, but the IR channel was able to detect thick and thin oil layers as well as the boundary between water and oil. Only in very favorable lighting and sea conditions were oil spills visible in the images [1, 22].

Another passive sensor is an MWR. The instrument is weather-insensitive because it only detects microwave radiation from the ocean in the cm to mm range. Oil slicks appear as bright objects on a darker sea because they emit microwave radiation that is stronger than the water. Oil slicks can have strong surface-emissivity signatures, but since determining the thickness of oil slicks requires a spatial resolution of tens to hundreds of meters, aircraft sensors are the most appropriate choice for this type of sensor for oil spill thickness monitoring [1, 6, 22].

Several studies for the monitoring and detection of oil spills compare SAR and optical sensors. For instance, the SAR data have the lowest backscatter levels in regions with algal blooms, whereas the SeaWiFS measures high levels of chlorophyll in these regions. Needless to say, different data sets could be used to distinguish between oil spills and look-alike phenomena like algal blooms [9, 22–24].

5. Potential of radar and microwave techniques for detecting oil spills

Oil on the sea surface dampens some of the small capillary waves that occur naturally in clean seas. Capillary waves reflect radar energy, producing a "bright" area in radar imagery. The presence of an oil slick can be detected as a "dark" area or one that has an absence of sea clutter. In this regard, Synthetic Aperture Radars (SARs) and Side-Looking Airborne Radars (SLARs) are the two primary types of radars used for environmental remote sensing and oil spill response. SLARs, despite being an older technology, are less expensive to purchase and use long antennas to improve along-track resolution. SARs achieve along-track resolution by using the forward motion of the sensor (an aircraft or spacecraft) to generate a very long antenna (which is range independent). A SAR signal requires sophisticated electronic processing to extract images. SARs are more expensive than SLARs but have a wider field of view and higher resolution. Comparative tests reveal that SAR is significantly better. Because they are made to identify complex targets, search, and rescue radar systems have little to no use for locating oil spills [16–22].

Radar detection of oil slicks is limited by sea state, low sea states will not produce sufficient clutter in the surrounding sea to contrast with the oil, and very high seas will scatter radar sufficiently to block detection inside the troughs. Indications are that wind speeds of at least 1.5 m/s are required as a minimum to allow detectability. Beyond this, wind speeds higher than 6 m/s will again remove the effect of an oil slick being distinguishable from the surrounding sea [9, 22–24].

Microwave radiation is emitted by the ocean. Because oil emits more microwave radiation than water, it appears as a "bright" area on a darker sea because oil is a stronger microwave emitter than water. Oil has a higher emissivity than water, which has a 0.4 emissivity factor. This difference in emissivity can be picked up by a passive device, which could act as a method of oil detection. The device could theoretically be used to measure the thickness of a slick because there is also a change in signal with thickness. This method has been very effective [20, 22, 24].

The methodology is dependent on prior knowledge of a range of environmental and oil-specific parameters, and the signal return is periodically influenced by oil thickness. Any one of two or three film thicknesses can be inferred from a given signal strength for a specified slick. When the effective thickness is an odd multiple of a quarter wavelength of the observed energy, microwave emission is at its highest. Additionally, the signal-to-noise ratio is low and biogenic materials interfere. Achieving high spatial resolution is challenging [1, 22, 24].

6. Look-alikes keystone issue in SAR data

The term frequently refers to an oil spill in the ocean. Oil spills can have devastating effects on the marine environment, including killing wildlife and polluting the water [6, 20, 23, 25]. In particular, an oil spill is a type of pollution that occurs when a liquid petroleum hydrocarbon is released into the environment as a result of human activity. Crude oil, refined petroleum products (like gasoline or diesel fuel) or byproducts, ship bunkers, oily waste, or waste mixed with oil are some of the different components that make up the oil. Pollution from the oil spill is challenging to remove. Natural oil seeps are another source of oil entering the marine environment. Although the majority of oil pollution caused by humans occurs on land, seagoing oil tankers have received the majority of public attention and regulatory attention. Unrelated to the oil spill, there are dark patches [9, 21, 22].

SAR satellite data is typically regarded as the most effective and superior satellite sensor for finding oil spills. Nevertheless, oil spill thickness estimation and oil type identification cannot be done with SAR data. The ability to distinguish between oil spills and look-alikes is the main issue with SAR data for oil spill detection. In actuality, both show up in SAR data as dark patches. Natural dark patches are referred to as "oil slick look-alikes" in this context. Natural films and slicks, ice, threshold wind speed regions (wind speeds of 3 m/s), wind protection from the land, rain cells, shear zones, internal waves, and other phenomena are examples of look-alikes (**Figure 6**). In a strict sense, an oil spill only refers to man-made slicks connected to crude petroleum and the products it produces, such as heavy and light fuel [9, 22, 24].

Currently, the SAR sensor is unable to differentiate between the various pollutants. However, for the large Sea Empress oil spill, there is a good correlation between the largest reduction in backscatter and the thickest oil as determined by visual

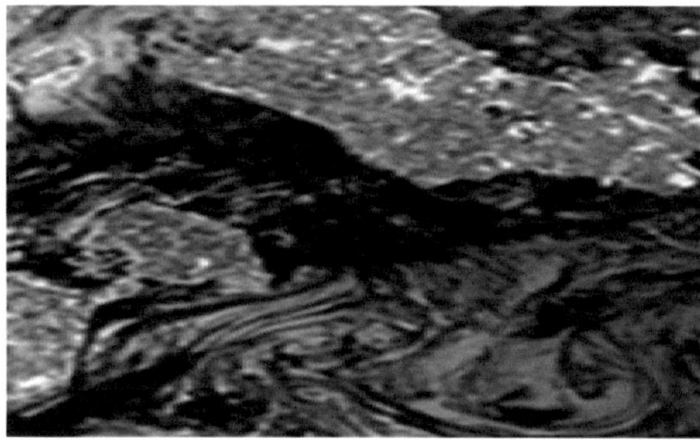

Figure 6.
Look-alike in SAR data.

observations for a constrained range of wind speeds (5–6 m/s). This suggests that a single SAR frequency may not be sufficient to estimate the thickness of the oil spill. When selecting features to distinguish between oil spills and lookalikes, these experiences must be taken into consideration. To distinguish between oil spills and look-alikes, physical, geometrical, and geographical parameters must be used, as well as significant characteristics like wind speed [9, 16–24].

7. Conclusion

Beyond oil spills, this chapter has demonstrated a variety of other concepts. It has looked into the effects of oil exploration and extraction on the environment, the economy, and politics, as well as the effects of spill-related harm on a region. As a result, the chapter also offers some fundamental knowledge about tracking oil spills from space. This is shown with the help of optical and microwave remote sensing technology.

The primary issue with using radar and microwave data to monitor an oil spill is the possibility of false alarms from look-alikes. As a consequence, there are a variety of issues that must be resolved regarding the detection of oil spills from space. Combining these technologies with other strategies, such as *in-situ* measurements and ground-based observations, is essential.

Author details

Maged Marghany
Global Geoinformation Sdn.Bhd., Kuala Lumpur, Malaysia

*Address all correspondence to: magedupm@hotmail.com

References

[1] Wang Z, Fingas M, Page DS. Oil spill identification. Journal of Chromatography A. 1999;**843**(1-2):369-411

[2] Takeshita R, Bursian SJ, Colegrove KM, Collier TK, Deak K, Dean KM, et al. A review of the toxicology of oil in vertebrates: What we have learned following the Deepwater horizon oil spill. Journal of Toxicology and Environmental Health, Part B. 2021;**24**(8):355-394

[3] Little DI, Sheppard SR, Hulme D. A perspective on oil spills: What we should have learned about global warming. Ocean & Coastal Management. 2021;**202**:105509

[4] Lourenço RA, Combi T, da Rosa AM, Sasaki ST, Zanardi-Lamardo E, Yogui GT. Mysterious oil spill along Brazil's northeast and southeast seaboard (2019-2020): Trying to find answers and filling data gaps. Marine Pollution Bulletin. 2020;**156**:111219

[5] Cheng H, Hu Y. Lead (Pb) isotopic fingerprinting and its applications in lead pollution studies in China: A review. Environmental Pollution. 2010;**158**(5):1134-1146

[6] Fingas M. The Basics of Oil Spill Cleanup. CRC Press; 2002

[7] Zhang W, Li C, Chen J, Wan Z, Shu Y, Song L, et al. Governance of global vessel-source marine oil spills: Characteristics and refreshed strategies. Ocean & Coastal Management. 2021;**213**:105874

[8] Doerffer JW. Oil Spill Response in the Marine Environment. Elsevier; 2013

[9] Otremba Z. The impact on the reflectance in VIS of a type of crude oil film floating on the water surface. Optics Express. 2000;**7**(3):129-134

[10] Marghany M. Genetic algorithm for oil spill automatic detection from ENVISAT satellite data. In: Computational Science and Its Applications–ICCSA 2013: 13th International Conference, Ho Chi Minh City, Vietnam, June 24-27, 2013, Proceedings, Part II 13. Berlin Heidelberg: Springer; 2013. pp. 587-598

[11] Etkin DS. Analysis of oil spill trends in the United States and worldwide. In International Oil Spill Conference 2001. 2001, 2, pp. 1291-1300). American Petroleum Institute

[12] Wang J, Zhou Y, Zhuang L, Shi L, Zhang S. Study on the critical factors and hot spots of crude oil tanker accidents. Ocean & Coastal Management. 2022;**217**:106010

[13] Stogiannidis E, Laane R. Source characterization of polycyclic aromatic hydrocarbons by using their molecular indices: An overview of possibilities. Reviews of Environmental Contamination and Toxicology. 2015;**234**:49-133

[14] Hoang AT, Pham VV, Nguyen DN. A report of oil spill recovery technologies. International Journal of Applied Engineering Research. 2018;**13**(7):4915-4928

[15] Reed M, Johansen Ø, Brandvik PJ, Daling P, Lewis A, Fiocco R, et al. Oil spill modeling towards the close of the 20th century: Overview of the state of the art. Spill Science & Technology Bulletin. 1999;**5**(1):3-16

[16] Fingas M, editor. Oil Spill Science and Technology. Gulf Professional Publishing; 2010

[17] Marghany M. RADARSAT for oil spill trajectory model. Environmental Modelling & Software. 2004;**19**(5):473-483

[18] Marghany M. Synthetic Aperture Radar Imaging Mechanism for Oil Spills. Gulf Professional Publishing; 2019

[19] Venkatesh S. The oil spill behaviour model of the Canadian atmospheric environment service part I: Theory and model evaluation. Atmosphere-Ocean. 1988;**26**(1):93-108

[20] Marghany M. Automatic Detection Algorithms of Oil Spill in Radar Images. CRC Press; 2019

[21] Marghany M. Radarsat Lagrangian flow effects on oil spill spreading. In: IEEE International Geoscience and Remote Sensing Symposium. Vol. 4. IEEE; 2002. pp. 2023-2025

[22] Brown CE, Fingas MF, Goodman RH. Oil-spill remote sensors: New tools that provide solutions to old problems. In: Arctic and Marine Oilspill Program Technical Seminar. Vol. 2. Canada: Ministry of Supply and Services; 1998. pp. 783-794

[23] Bello J, Eriksen P, Pocwiardowski P. Oil leak detections with a combined telescopic fluorescence sensor and a wide band multibeam sonar. In: International Oil Spill Conference Proceedings. Vol. 1. International Oil Spill Conference; 2017. pp. 1559-1573

[24] Marghany M. Finite element model of residual currents and oil spills transport. InIGARSS 2000. In: IEEE 2000 International Geoscience and Remote Sensing Symposium. Taking the Pulse of the Planet: The Role of Remote Sensing in Managing the Environment. Proceedings (cat. No. 00CH37120). Vol. 7. IEEE; 2000. pp. 2966-2968

[25] Topouzelis KN. Oil spill detection by SAR images: Dark formation detection, feature extraction and classification algorithms. Sensors. 2008;**8**(10):6642-6659

Chapter 2

Perspective Chapter: Understanding Thermal Maturity Evolution and Hydrocarbon Cracking – Implication for Cretaceous Awgu and Nkporo Shales, Southeastern Nigeria

Jerry Danwazan, Adesina Mutiu Adeleye, Trust-Paul Bakyu, Mamidak M. Iliya and Joseph Kehinde Egunjobi

Abstract

One-dimensional basin modeling was carried out using Schlumberger's PetroMod modeling software that provided understanding on the thermal evolution, timing of hydrocarbon generation and expulsion of the Coniacian Awgu Shale and the Campanian Nkporo Shale penetrated by Nzam-1 and Akukwa-2 wells in the lower Benue Trough, Nigeria. The burial temperature and vitrinite reflectance values ranged from 30 to 145°C and 0.5 to 2.9%Ro for Awgu Formation, 28 to 125°C and 0.5 to 1.5%Ro for Nkporo Formation in Nzam-1 well model; 29.5 to 145°C and 0.8 to 2.4%Ro for Awgu Formation, and 28.5 to 95°C and 0.6 to 0.8%Ro for Nkporo Formation in Akukwa-2 well model. Awgu Shale reached the required threshold of the oil generation window during mid Campanian (75Ma) and late Santonian (82Ma) in Nzam-1 and Akukwa-2 well models, respectively. Nkporo Shale entered the required oil window threshold during early Paleocene (65Ma) in Nzam-1 well model and late Maastrichtian (67Ma) in Akukwa-2 well model. This study revealed that valid petroleum system elements exist in Anambra basin, and some amount of gaseous hydrocarbons and little oil may have been generated and expelled. Exponential decrease in temperature over time has favored the preservation of the gas reservoirs and the survival of hydrocarbons in the deep strata. The early maturity of Nkporo Shale can be attributed to lack of the requisite burial depth, temperature and pressure in favor of oil generation and expulsion. Post-maturity status of Awgu Shales may be associated with deeper burial depth and possibly due to the effect of Santonian tectonic episode.

Keywords: one dimensional modeling, transformation ratio, thermal maturity, hydrocarbon generation, Awgu and Nkporo Shales

1. Introduction

Petroleum (oil and gas) accounts for up to 95% of the Nigeria's foreign earnings and has remained the major supporter of its economy since it was first discovered in commercial quantity in 1956. Globally, petroleum as an energy source will continue to dominate other primary energy sources and is expected to account for up to 56% of the world energy demand in the year 2030 [1]. This makes source rock studies and evaluation now to be of key interest to oil industry players, the Academia and other research interest groups and as a result research has also been intensified in the Lower Benue Trough as well.

The thermal states of sedimentary basins affect hydrocarbon generation, migration, and accumulation processes, and therefore reconstructing a basin's thermal history is significant for petroleum accumulation analyses. The maturity of organic matter is primarily controlled by the temperature of the source rock [2].

We are interested in thermal evolution and hydrocarbon cracking because of its importance in modeling the oil degradation processes that compete with oil expulsion during maturation of petroleum source rock over geologic time periods.

As a source rock begins to mature, it generates hydrocarbons. As an oil-prone source rock matures, the generation of heavy oil is succeeded by medium and light oils and condensates. Above a temperature of approximately 100°C, only dry gas is generated and incipient metamorphism is imminent. The maturity of a source rock reflects the ambient temperature as the conditions favorable for hydrocarbon generation. Understanding the maturation and thermal cracking of hydrocarbon helps to define the quality of the expected products.

Following the increased interest in the exploration of petroleum resources, Assessment of generative potential and characteristics of source rocks is fundamental in hydrocarbon exploration and its success depends largely on the employed organic geochemical method and this calls for a more refined and integrated approach by both the industry and the academia so as to discover more hydrocarbon prospects and despite the fact that published data gives a vast amount of information on the geology, sedimentology, lithostratigraphy and the hydrocarbon generation potentials of the Lower Benue Trough yet of the several Geochemical approaches for source rock evaluation employed in the studied area by some of the existing works which include using vitrinite as a maturity tool; there are only but a few among these previous researches that integrated the other source rock evaluation techniques with burial history and Maturity Modelling which could have given a clearer and better understanding of the thermal evolution, hydrocarbon generation and timing of the organic matter (kerogen) in this section (Nkporo and Agwu Formations) of the Lower Benue Trough.

Ehinola [3] investigated and presented results on subsurface geological models of the Anambra basin using the Petromod Software. The models produced were based on a clear understanding of how sedimentary rocks are formed, modified and perform as source rocks, reservoirs and traps employing lithostratigraphic principles, rock mineralogy and depositional modeling. The study revealed that thermal maturity of the source rock units in Anambra basin increases toward the Abakaliki Anticlinorium and with depth towards the western part of the basin. Imo shale, Mamu Formation, Nkporo Shale, Awgu Shale, Eze-Aku shale and Asu River group reached hydrocarbon generation at 51.23, 61.42, 56.30, 66.11,84 and 92Ma respectively which show that the Asu River Group and Ezeku-Aku Formation shale commenced Hydrocarbon generation prior to the Santonian tectonism. This maybe reflected by the presence of seepages, oil and gas in the Owelli, Agwu and Eze-Aku Formation interval respectively in the Anambra Basin while the matured Post Santonian succession would generate and charge both the upper Cretaceous reservoir and possibly the sub-Niger Delta successions in the subsurface. The study also revealed that the Maastrichtian-Paleocene sediment in the South-western part of the basin maybe considered as the major strata with liquid hydrocarbon potential whereas the Pre-Santonian sediment exhibit potential for gaseous hydrocarbon.

Oluwajana [4] carried out a hydrocarbon-charge modeling and its consequence on shale gas and shale oil resource systems within the Cretaceous strata of the Anambra Basin using organic geochemical data, 2D seismic line and stratigraphic well data of exploratory wells drilled in the basin to generate conceptual models, improve understanding on hydrocarbon generation and timing, and identify potential shale plays in the basin and the study helped to identify potential shale resource system in the Anambra basin namely the upper Cenomanian–lower Turonian marine facies.
Two main shale play types involving the upper Cenomanian–lower Turonian were identified.

Akaegbobi et al. [5] revealed that the organic matter in majority of Nkporo samples and few of Owelli samples can be classified as predominantly Type II-III kerogen, while the organic matter in the rest of Owelli and Mamu formation samples is majorly Type III and Type IV kerogens and further revealed that the analyzed samples were deposited in normal marine environment with samples from Mamu Formation tending towards freshwater depositional millieux".

Akaegbobi et al. [5] investigated the shales, siltstones and shale heteroliths, collected from the Nkporo, Owelli and Mamu formations of Anambra Basin were subjected to bulk and molecular geochemical analyses (total organic carbon content determination, pyrolysis analysis, bitumen extraction and gas chromatography) to provide further insight on the quantity, quality and thermal maturity of organic matter within the sediments, the source input and paleodepositional conditions of the organic matter, hydrocarbon generation potential and the study revealed that the organic matter has fair to good hydrocarbon generation potential and the majority of the analyzed sediments from Nkporo Formation are oil and gas prone, while the others are mainly gas prone. The Molecular geochemical data further suggested that the organic matter within the sediments was derived from mixed aquatic algae and land plant source input and was deposited under suboxic paleodepositional conditions.

Adeleye et al. [6] attempted to evaluate aspects of the source rock potential for hydrocarbon generation of the Imo Shale Formation penetrated by the Akukwa II and Nzam-Iwells in Anambra basin, and concluded that the sediments have potential for hydrocarbon generation with possibility of gas. Organic matter contained in the sediments is predominantly type IV kerogen sourced from terrestrial materials which does not yield significant amounts of hydrocarbon. Thermal maturity derived from Rock-eval data revealed that the Imo Formation samples are immature with respect to hydrocarbon generation.

Adeleye et al. [7] employed Total Organic Carbon (TOC) content and Rock-eval Pyrolysis to evaluate source rock potential for hydrocarbon generation of Nkporo Formation in the Lower Benue trough penetrated by Nzam-1 well and it was revealed that the sediments contain poor to fair source rock for hydrocarbon with kerogen type III as the predominating organic matter, which is capable of generating dry gas. Tmax and other pyrolysis data suggest that the organic matter in the Nkporo Formation is at the peak of thermal maturity to post maturity with respect to hydrocarbon generation.

And the study revealed that the heat energy generated from post mature part of the studied section together with the thermal maturity peak to late maturity generally observed for the sediments may have resulted in the dry gas prospect.

Adebayo et al. [8] carried out a Palynological, organic petrographic, and organic geochemical analyses of the Campanian-Maastrichtian sediments in Akukwa-2 well to infer their paleoenvironments, origin of the organic matter, and hydrocarbon generation potentials and the study revealed that the organic matter within the sediments is also likely to generate mainly gas. This is in agreement with the petrographic observations, which revealed that the analyzed shale samples contain abundant vitrinite macerals, apart from bituminite, alginite, cutinite, and resinite. Also, the sediments are immature to early mature in terms of hydrocarbon generation as indicated by vitrinite reflectance, biomarker maturity, and pyrolysis Tmax data. Biomarker distribution ratios, palynomorphs assemblage, and organic petrographic observations further point out that the organic materials within the sediments were of mixed aquatic and terrigenous origin and were deposited under suboxic paleodepositional conditions. Based on sedimentological, palynological, and biomarker characteristics, the environment of deposition of the analyzed sediments was inferred to be a relatively quiet, shallow marine with fluvial incursion, most especially at the upper part of the intervals studied and consequently, it is a delta associated depositional environment with a fluviatile influence. The sediments were therefore suggested to be deposited in a paleogeographic setting close to vegetation source.

Following the increased interest in the exploration of petroleum resources, Assessment of generative potential and characteristics of source rocks is fundamental in hydrocarbon exploration and its success depends largely on the employed organic geochemical method and this calls for a more refined and integrated approach by both the industry and the academia so as to discover more hydrocarbon prospects and despite the fact that published data gives a vast amount of information on the geology, sedimentology, lithostratigraphy and the hydrocarbon generation potentials of the Lower Benue Trough yet of the several Geochemical approaches for source rock evaluation employed in the studied area by some of the existing works which include using vitrinite as a maturity tool; there are only but a few among these previous researches that integrated the other source rock evaluation techniques with burial history and Maturity Modelling which could have given a clearer and better understanding of the thermal evolution, hydrocarbon generation and timing of the organic matter (kerogen) in this section (Nkporo and Agwu Formations) of the Lower Benue Trough.

Therefore, this current study attempts to evaluate the characteristics of the source rocks and their viability, hydrocarbon generation potential and timing, and predict the various thermal maturity levels of the possible shale plays in the Coniacian Agwu Formation and late Campanian Nkporo Formation sediments within this section of the Lower Benue Trough as penetrated by the Nzam-1 well and Akukwa-2 wells respectively. A quantitative one dimensional basin modeling was carried out for evaluating the thermal histories and timing of hydrocarbon generation and expulsion of the Coniacian and late Campanian source rocks in the part of Lower Benue Trough. More so, the reconstruction of the burial, thermal and maturity histories were modeled in order to evaluate the remaining hydrocarbon potential using Schlumberger's PetroMod (1D) modeling software so as to provide the basis for Petroleum resource evaluation of the Upper Cretaceous sediments of the basin. Also, in this study, a detail evaluation of the acquired TOC and Rock-Eval pyrolysis data was carried out which provided information on the quantity, quality and maturity of the organic matter which served as the verification for the results and interpretations from the model.

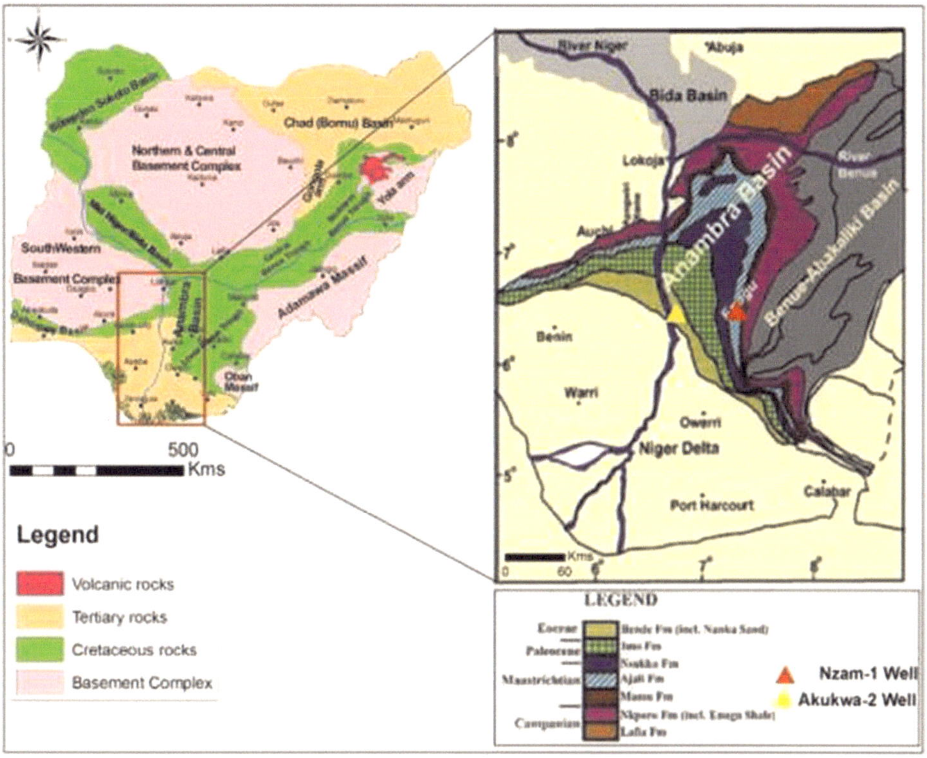

Figure 1.
A Geological map of Nigeria showing the location of the studied exploratory well, Nzam-1 [8].

The study area is carved out of the Lower Benue Trough (**Figure 1**), in South Eastern Nigeria as penetrated by the Exploratory wells, Nzam-1 and Akukwa-2. The area comprises the geographical location of the following states: Abakaliki, Anambra, Ebonyi, Enugu, and Imo and lies within Latitude N 6°27′17.07″ and Longitudes E6° 43′10.75″ on the Geological map of Nigeria employing the ArcGIS Software (**Figure 1**).

2. Aim and objectives of study

This chapter is aimed at reconstructing the burial and thermal history of the organic matter in the Anambra Basin as it affects the maturity and thermal cracking of the Coniacian Awgu Shales and the late Campanian Nkporo Shales through the thermal evolution conditions for oil and natural gas accumulation.

Therefore, this chapter attempts to achieve the following objectives:

i. To generate one dimensional burial/thermal history model of the two wells (Nzam-1 and Akukwa-2) which penetrated the Coniacian-late Campanian Source rocks in the Anambra Basin

ii. To determine the thermal regime changes during hydrocarbon generation and preservation

iii. To determine organic matter evolution, timing of hydrocarbon generation/ cracking, accumulation and expulsion

iv. To Identify the reservoirs and the possible seal through the interpretation of the Petroleum system elements (PSE)Models

v. To utilize the deductions from this study to provide information necessary to optimize exploration activities in the Lower Benue Trough with a view of improving past investigation in the study area.

3. Methodology

Geohistory diagrams and similar diagrams have been widely used in geology, particularly in hydrocarbon exploration. These diagrams were adapted to perform numerical modeling of burial, erosion, and thermal histories in sedimentary basins, e.g. [9, 10].

In this study, quantitative one dimensional basin modeling (1-D) is performed for evaluating the thermal histories and timing of hydrocarbon generation and expulsion of the Agwu and Nkporo source rocks in the Lower Benue Trough. The 1-D Basin modeling, was used for the reconstruction of the burial and temperature histories of the two studied wells (**Figure 2**). The reconstruction of the burial, thermal and maturity histories were modeled in order to evaluate the remaining hydrocarbon potential using Schlumberger's PetroMod (1D) modeling software and Lithology data comprising of sandstone, shale and absolute ages.

The modeling approach adopted requires input data which describe the present day geological situation as a result of past events (**Figure 3**). On this basis, the geological history is simulated from the oldest to the most recent. This also involves modeling and calibration of single-point data, whereby geologic and geochemical information are integrated to model the Formations and evolution of a sedimentary basin. As a stand-alone tool, the single point data (well) is constructed and imported from the well editor. The total sedimentary column (lithology) was determined based on well data. Erosion and heat flow changes were established through calibration against maturity data, giving rise to the generation of a conceptual model of the subsidence and thermal history of the region. This model of present day architecture represents the final result of all the processes acting on the basin throughout geologic time. Summarily, the following input data are required for reconstruction of burial history:

i. Depositional thickness

ii. Depositional age in Ma (millions of years)

iii. Lithological Composition

iv. Thickness and age of eroded intervals, Petroleum Systems Essential Elements (Underburden, Source Rocks, Reservoir rocks, Seal Rock and Overburden rock

v. Possible source rock properties (TOC & HI).

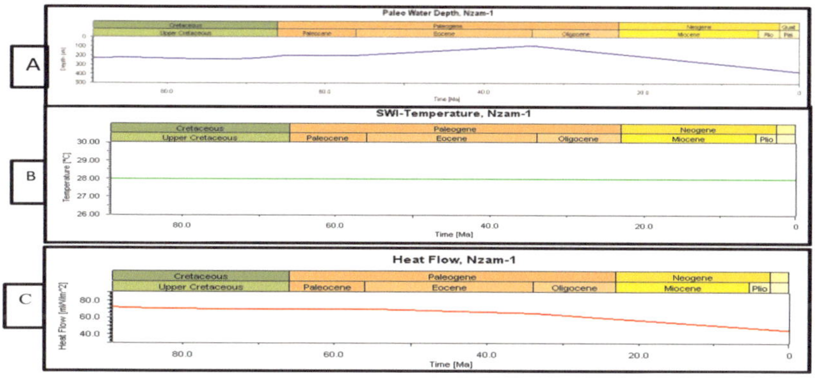

Figure 2.
Plots of Boundary conditions for the studied well (Nzam-1) ; (A) Paleo Water depth Versus geologic Age in million years, notice its fairly constant value of about 240 m between upper Cretaceous and Paleocene and between late Eocene and early Oligocene witnessed to about 50 m and its rapid increase by Oligocene to present day value of about 400 m (B) Sediment Water Interface-Temperature versus geologic Age in Million years, notice the fairly constant value of 28°C (C) Heat flow versus geologic Age in Millions years, notice maximum value of 72 m/Wm^2 between upper Cretaceous and Paleocene and a constant decline from Eocene to a present day value of 48 m/Wm^2.

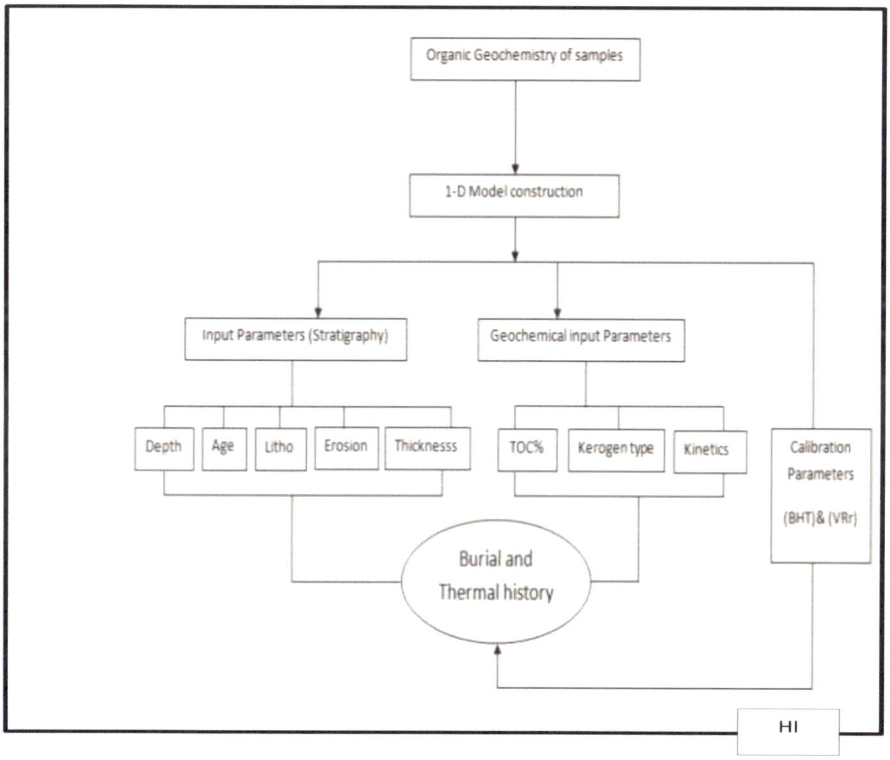

Figure 3.
Flow chart indicating the steps involved in the construction of the one dimensional burial/thermal history Modeling.

The geologic model consisting of the depositional, non-depositional and erosional events, was compiled using stratigraphic data which were provided from well reports, bank of Shell Petroleum Development Company (SPDC) of Nigeria and previous stratigraphic studies (**Figures 4** and **5**). Hydrocarbon generation modeling was based on TOC and HI of the Coniacian Awgu and the upper Campanian Nkporo source rocks in the Lower Benue Trough. The modeling results were also calibrated with computed vitrinite reflectance after [11] and borehole temperatures (BHT) in the study area. Geochemical data (TOC and pyrolysis data), Vitrinite reflectance (**Tables 1–3**), Paleo water depth, well log, and geologic data (**Tables 4** and **5**) were used for the construction of the geologic model showing thermal maturity stages of the Agwu and Nkporo Source rocks in the study area. Lithology data comprising of sandstone, shale and absolute ages were put into use (**Figures 4** and **5**). The hydrocarbon generation modeling was based on TOC and HI of the Agwu and Nkporo source rocks in the Lower Benue Trough. The maturity modeling was calculated using the EASY% Ro model of [14, 15]. The TOC and Pyrolysis data of Coniacian Agwu Shales in Nzam-1 were extracted from [12] while those of the upper Campanian Nkporo Shales in Nzam-1 were extracted from [7]. TOC and Pyrolysis data for upper Campanian Nkporo Source Rock in Akukwa-2 were extracted from [8] and the TOC and HI values for Coniacian Awgu Source Rock in Akukwa-2 were extracted from Abubakar [1] (**Figure 6**).

The Modelled Vitrinite Reflectance as calculated after [14] has been related to calculated vitrinite reflectance data after [11] so as to enable the calibration of thermal history (**Figures 7** and **8**).

Paleo-water depth values were used to define the paleogeometry while heat flow and sediment water interface temperatures were the key boundaries conditions that were defined in the course of modeling. Thermal evolution is simulated on the basis of boundary assignments applied to certain time steps. The assigned parameters are heat flow densities in mW/m^2 and surface temperatures in °C. Acquired Total Organic Carbon Content (TOC) and Rock eval pyrolysis of the Agwu and Nkporo Shales from Nzam and Akukwa-2 wells in the Lower Benue Trough are presented here with some additions in respect of some calculated pyrolysis parameters for the purpose of this study. Acquiring these data became necessary because the values of

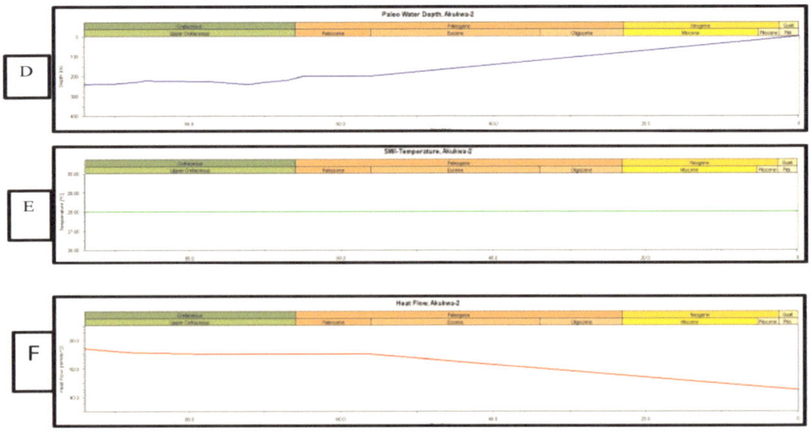

Figure 4.
Plots of Boundary conditions for the studied well (Akukwa-2); (D) Paleo Water depth Versus geologic Age in million years, notice its fairly constant value of about 240 m between Upper Cretaceous and Paleocene and between late Eocene and early Oligocene witnessed decline to about 50 m and its rapid decrease by Oligocene to present day value of about 0 m (E) Sediment Water Interface-Temperature versus geologic Age in Million years, notice the fairly constant value of 28°C (F) Heat flow versus geologic Age in Millions years, notice maximum value of 72 m/Wm² between upper Cretaceous and Paleocene and a constant decline from Eocene to a present day value of 48 m/Wm².

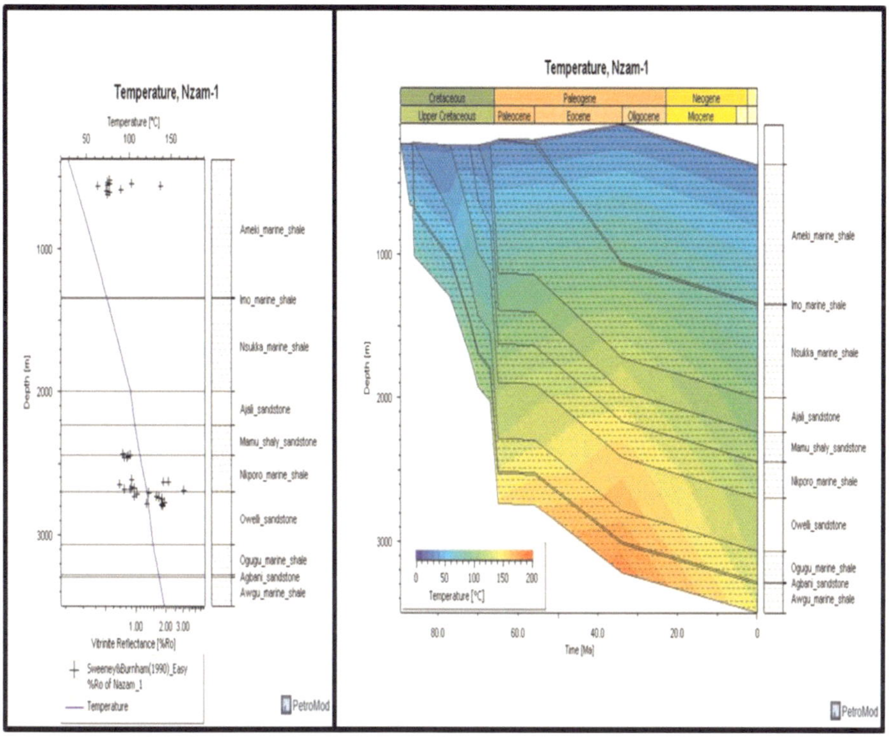

Figure 5.
Paleo-temperature modeling in Nzam-1 well calibrated using borehole temperature; showing correlation among Burial history with temperature overlay, measured temperature and modeled Temperature for the studied well, notice that the maximum temperature values of 120°C–145°C in the area was attained between mid-Paleocene and mid Miocene (60-15ma) on Coniacian Agwu Source strata, higher temperatures are associated with Santonian tectonic episode

Hydrogen Index (HI), TOC, Vitrinite reflectance forms vital input parameters in the one dimensional burial/thermal history model construction.

4. Results from 1-D burial/thermal history modeling

The input parameters used for constraining 1-D models/plots constructed for the studied sections of Nzam-1 and Akukwa-2 wells in the Lower Benue Trough are shown in **Tables 4** and **5** and the outcomes of the 1-D models/plots constructed for the studied sections of Nzam-1 and Akukwa-2 wells in the Anambra Basin, Lower Benue Trough are shown in **Figures 3–7, 9–12**.

5. Discussion

5.1 Organic matter evolution, thermal regime changes during hydrocarbon generation, preservation and expulsion

Figures 2 and **4** illustrate that the end of the Cretaceous era saw uplift, denudation, subsidence and basin cooling in the Lower Benue Trough as well as increased

Depth (m)	Formation	TOC wt.%	S1 (Mg HC/ g TOC)	S2 (Mg HC/ g TOC)	S4 (Mg HC/ g TOC)	S1+S2 (Mg HC/ g TOC)	Tmax (°C)	Calc %Ro	*Calc %Ro	%Ro Diff.	HI Mg HC/ g TOC	PI	(%) PC	OSI (Mg HC/ g TOC)
2467	Nkporo	1.12	0.16	0.24	10.9	0.4	439	0.74	0.69	0.05	21	0.4	0.03	14.29
2477	Nkporo	0.99	0.14	0.24	9.54	0.38	446	0.87	0.80	0.07	24	0.37	0.03	14.14
2482	Nkporo	1.03	0.21	0.38	9.83	0.59	444	0.83	0.77	0.06	37	0.36	0.05	20.39
2487	Nkporo	1.09	0.22	0.36	10.43	0.58	442	0.81	0.74	0.07	33	0.38	0.05	20.18
2492	Nkporo	1.07	0.18	0.24	10.37	0.42	440	0.76	0.71	0.05	22	0.43	0.03	15.82
2647	Nkporo	1.16	0.17	0.31	11.23	0.48	448	0.90	0.83	0.083	27	0.35	0.04	14.66
2652	Nkporo	0.08	NA	NA	NA	NA	NA	NA	NA	NA	NA	NA	NA	NA
2657	Nkporo	1.4	0.21	0.31	13.57	0.52	516	2.13	1.84	0.29	22	0.4	0.04	15.00
2662	Nkporo	1.25	0.18	0.25	12.11	0.43	503	1.89	1.64	0.25	20	0.42	0.04	14.40
2677	Nkporo	1.14	0.14	0.25	11.05	0.39	436	0.69	0.65	0.04	22	0.36	0.03	12.28
2692	Nkporo	1.08	0.15	0.23	10.53	0.38	448	0.90	0.83	0.07	21	0.39	0.03	13.89
2702	Nkporo	1.12	0.13	0.24	10.93	0.37	451	0.96	0.87	0.09	21	0.35	0.03	11.61
2707	Nkporo	1.03	0.13	0.24	9.99	0.37	446	0.87	0.80	0.07	23	0.35	0.03	12.62
2712	Nkporo	1.45	0.12	0.2	14.2	0.32	440	0.76	0.71	0.05	14	0.38	0.03	8.28
2717	Nkporo	0.08	NA	NA	NA	NA	NA	NA	NA	NA	NA	NA	NA	NA

Table 1.
Acquired TOC content and pyrolysis data with calculated parameters of the studied samples of Nkporo Formation, Nzam-1 well [7].

Sample I.D	Depth (m)	Formation	TOC wt.%	S1 (Mg HC/ g TOC)	S2 (Mg HC/ g TOC)	S3 (Mg HC/ g TOC)	S4 (Mg HC/ g TOC)	S1+S2 (Mg HC/ g TOC)	Tmax (°C)	Calc %Ro	*Calc %Ro	% Ro Diff	HI (Mg HC/ g TOC)	OI (Mg HC/ g TOC)	PI	PC (%)	OSI (Mg HC/ g TOC)
A16	2722	Agwu	1.13	0	0.01	2.26	11.34	0.01	568	3.06	2.61	0.45	1	200.39	0	0	0.00
A17	2737	Agwu	1.27	0.19	0.32	2.51	12.28	0.51	470	1.33	1.15	0.18	25	197.73	0.37	0.04	14.96
A18	2742	Agwu	0.44	NA	NA	NA	NA	NA	NA	NA	NA	NA	NA	NA	NA	NA	NA
A19	2747	Agwu	1.39	0.19	0.3	2.76	13.5	0.49	454	1.01	0.91	0.10	22	198.34	0.39	0.04	13.67
A20	2757	Agwu	1.45	0.21	0.32	2.89	14.05	0.53	486	1.59	1.39	0.20	22	`199.30	0.4	0.04	14.48
A21	2762	Agwu	1.33	0.17	0.28	2.65	12.92	0.45	451	0.96	0.87	0.09	21	199.56	0.38	0.04	12.78
A22	2772	Agwu	1.42	0.16	0.27	2.85	13.83	0.43	493	1.71	1.50	0.21	19	200.59	0.37	0.04	11.27
A23	2777	Agwu	1.4	0.19	0.25	2.8	13.63	0.44	499	1.82	1.59	0.23	18	199.82	0.43	0.04	13.57
A24	2803	Agwu	1.78	0.2	0.25	3.56	17.44	0.45	505	1.93	1.67	0.26	14	199.76	0.44	0.04	11.24
A25	2808	Agwu	1.44	0.19	0.24	2.85	14.07	0.43	469	1.28	1.14	0.14	17	197.87	0.44	0.04	13.19
A26	2813	Agwu	1.25	0.14	0.17	2.48	12.27	0.31	500	1.84	1.60	0.24	14	198.42	0.45	0	11.20
A27	2818	Agwu	1.38	0.16	0.22	2.8	13.45	0.38	501	1.86	1.61	0.25	16	202.65	0.42	0.03	11.59
A28	2823	Agwu	1.31	0.15	0.2	2.66	12.77	0.35	497	1.79	1.56	0.23	15	203.29	0.43	0.03	11.45
A29	2828	Agwu	1.4	0.22	0.23	2.83	13.59	0.45	501	1.86	1.61	0.25	16	201.96	0.49	0.04	15.71
A30	2833	Agwu	0.07	NA	NA	NA	NA	NA	NA	NA	NA	NA	NA	NA	NA	NA	NA

Table 2.
Acquired TOC content and pyrolysis data with calculated parameters of the studied samples of Agwu Formation, Nzam-1 well (modified after [12]).

Depth (m)	Formation	TOC (Wt. %)	S1 (Mg HC/ g TOC	S2 (Mg HC/ g TOC)	Tmax (°C)	Ro	Calc (% Ro)	*Calc (% Ro)	%Ro Diff.	HI (Mg HC/ g TOC)	PI	OSI (Mg HC/ g TOC)
1348	Nkporo	0.89	0.1	1.44	430	N/A	0.58	0.56	0.02	162	0.06	11.24
1351	Nkporo	0.81	0.14	0.82	434	N/A	0.65	0.62	0.03	101	0.15	17.28
1354	Nkporo	1.27	0.22	1.21	431	N/A	0.60	0.57	0.03	95	0.15	17.32
1357	Nkporo	0.68	0.14	0.93	433	N/A	0.63	0.60	0.03	137	0.13	20.59
1360	Nkporo	0.63	0.13	1.13	430	N/A	0.58	0.56	0.02	179	0.10	20.63
1373	Nkporo	1.8	0.41	1.36	427	0.56	0.53	0.51	0.02	76	0.23	22.78
1376	Nkporo	1.76	0.22	1.16	431	N/A	0.60	0.57	0.03	66	0.16	12.50
1379	Nkporo	1.88	0.26	1.73	430	N/A	0.58	0.56	0.02	92	0.13	13.83
1400	Nkporo	0.27	0.16	0.47	431	0.56	0.60	0.57	0.03	174	0.25	59.26
1410	Nkporo	1.92	0.15	1.45	428	N/A	0.54	0.53	0.01	76	0.09	7.81
1425	Nkporo	1.5	0.13	1.05	432	0.55	0.62	0.59	0.03	70	0.11	8.67
1428	Nkporo	1.52	0.2	1.2	428	N/A	0.54	0.53	0.01	79	0.14	13.16
1443	Nkporo	1.29	0.09	0.89	433	N/A	0.63	0.60	0.03	69	0.09	6.98
1467	Nkporo	1.07	0.05	0.93	429	N/A	0.56	0.54	0.02	87	0.05	4.67
1473	Nkporo	1.43	0.14	1.1	434	N/A	0.65	0.62	0.03	77	0.11	9.79
1482	Nkporo	1.16	0.17	0.95	433	N/A	0.63	0.60	0.03	82	0.15	14.66
1485	Nkporo	0.98	0.1	1.01	431	N/A	0.60	0.57	0.03	103	0.09	10.20
1492	Nkporo	1.67	0.14	1.24	431	N/A	0.60	0.57	0.03	74	0.10	8.38
1495	Nkporo	1.56	0.18	0.55	434	N/A	0.65	0.62	0.03	35	0.25	11.54
1498	Nkporo	0.4	0.07	0.65	433	N/A	0.63	0.60	0.03	163	0.10	17.50
1504	Nkporo	1.43	0.16	1.19	430	N/A	0.58	0.56	0.02	83	0.12	11.19
1513	Nkporo	0.95	0.15	1.61	431	N/A	0.60	0.57	0.03	169	0.09	15.79
1516	Nkporo	3.02	0.13	1.19	431	N/A	0.60	0.57	0.03	39	0.10	4.30
1525	Nkporo	1.42	0.07	1.02	429	N/A	0.56	0.54	0.02	72	0.06	4.93
1535	Nkporo	1.84	0.05	1.06	433	0.6	0.63	0.60	0.03	58	0.05	2.72
1541	Nkporo	1.62	0.12	1.36	431	0.57	0.60	0.57	0.03	84	0.08	7.41
1589	Nkporo	1.48	0.07	1.23	433	0.58	0.63	0.60	0.03	83	0.05	4.73
1632	Nkporo	2.3	0.13	2.36	437	0.57	0.71	0.66	0.05	103	0.05	5.65
1663	Nkporo	1.95	0.18	1.82	434	N/A	0.65	0.62	0.03	93	0.09	9.23
1702	Nkporo	2.31	0.21	3.14	441	0.6	0.78	0.72	0.06	136	0.06	9.09

Table 3.
Acquired TOC content and pyrolysis data with calculated parameters of the studied samples of Nkporo Shales in Akukwa-2 well (modified after [7]).

geothermal gradient to have caused complex processes of uplift, denudation, and basin cooling and heating (**Figure 13**).

The geology of Lower Benue Trough is associated with the tectonic activities that were recorded during the Cenomanian which produced uplift with a NE-SW trend

Formation	Age in millions years	Heat flow (mw/m^2)	Eustatic sea level (m) [13]
	0	45	52
Ameki	56.0–33.9	70–65	201
Imo	65–56.0	70	200
Nsukka	67.0–65.0	70	220
Ajali	70.0–67.0	70	238
Mamu	72.1–70	70	241
Nkporo	**77.3–72.1**	**70**	**240**
Owelli	80.0–77.3	71	227
(Non-deposition)	86.0–80.0	71	222
Ogugu	86.3–86.0	71	225
Agbani	87.0–86.3	71	228
Awgu	**89.8–87.0**	**72**	**`237**

Table 4.
Input Parameters (age, Heat flow and Eustatic Sea level) of Lower Benue Trough used to constrain the model.

Age [Ma]	Name top/well pick	Depth [m]	Thickness [m]	Event type	Name layer/event	Paleodeposition/ erosion [m]	Lithology	PSE	Kinetic	TOC [%]	HI [mgHC/gTOC]
33.90	Ameki	375									
			963	↓ Deposition	Ameki_marine_shale		Shale (black)	Seal Rock			
56.00	Imo	1338									
			15	↓ Deposition	Imo_marine_shale		Shale (black)	Seal Rock			
65.00	Nsukka	1353									
			647	↓ Deposition	Nsukka_marine_shale		Shale (black)	none			
67.00	Ajali	2000									
			237	↓ Deposition	Ajali_sandstone		Sandstone (typical)	Reservoir Rock			
70.00	Mamu	2237									
			209	↓ Deposition	Mamu_shaly_sandstone		Shale (organic lean, sandy)	none			
72.10	Nkporo	2446									
			248	↓ Deposition	Nkporo_marine_shale		Shale (organic rich, typical)	Source Rock	Vandenbroucke_et_al(1999)_TIII-(NorthSea)	1.45	37.00
77.00	Owelli	2694									
			372	↓ Deposition	Owelli_sandstone		Sandstone (typical)	Reservoir Rock			
86.00	Ogugu	3066									
			214	↓ Deposition	Ogugu_marine_shale		Shale (organic rich, typical)	Source Rock			
86.30	Agbani	3280									
			15	↓ Deposition	Agbani_sandstone		Sandstone (typical)	Reservoir Rock			
87.00	Awgu	3295									
			200	↓ Deposition	Awgu_marine_shale		Shale (organic rich, typical)	Source Rock	Vandenbroucke_et_al(1999)_TIII-(NorthSea)	1.78	25.00
89.30	Awgu_Base	3495									

(A)

Age [Ma]	Name top/well pick	Depth [m]	Thickness [m]	Event type	Name layer/event	Paleodeposition/ erosion [m]	Lithology	PSE	Kinetic	TOC [%]	HI [mgHC/gTOC]
56.00	Imo	0									
			284	↓ Deposition	Imo_Marine shale		Shale (black)	Seal Rock			
65.00	Nsukka	284									
			225	↓ Deposition	Nsukka_Marine shale		Shale (black)	Seal Rock			
67.00	Ajali	509									
			430	↓ Deposition	Ajali_sandstone		Sandstone (typical)	Reservoir Rock			
70.00	Mamu	939									
			332	↓ Deposition	Mamu_shaly-sandstone		Shale (organic lean, sandy)				
72.10	Nkporo	1271									
			204	↓ Deposition	Nkporo_Marine shale		Shale (organic rich, typical)	Source Rock	Vandenbroucke_et_al(1999)_TIII-(NorthSea)	1.92	179.00
77.00	Owelli	1475									
			174	↓ Deposition	Owelli_sandstone		Sandstone (typical)	Reservoir Rock			
86.00	Ogugu	1649									
			924	↓ Deposition	Ogugu_Marine shale		Shale (organic rich, typical)	Source Rock			
86.30	Agbani	2573									
			198	↓ Deposition	Agbani_sandstone		Sandstone (typical)	Reservoir Rock			
87.00	Awgu	2771									
			640	↓ Deposition	Awgu_Marine shale		Shale (organic rich, typical)	Source Rock	Vandenbroucke_et_al(1999)_TIII-(NorthSea)	1.88	174.00
89.80	Eze-Aku	3411									
			90	↓ Deposition	Eze-Aku_Marine shale		Shale (organic rich, typical)	Source Rock			
93.90	Eze-Aku-Base	3501									

(B)

Table 5.
Inputed data (Lithology, Geologic age in Million Years (Ma), Petroleum system elements(PSE), TOC) for the burial history and maturity model construction of Coniacian and late Campanian source rocks, Lower Benue Trough, Nzam-1 (A) and Akukwa-2 (B) wells.

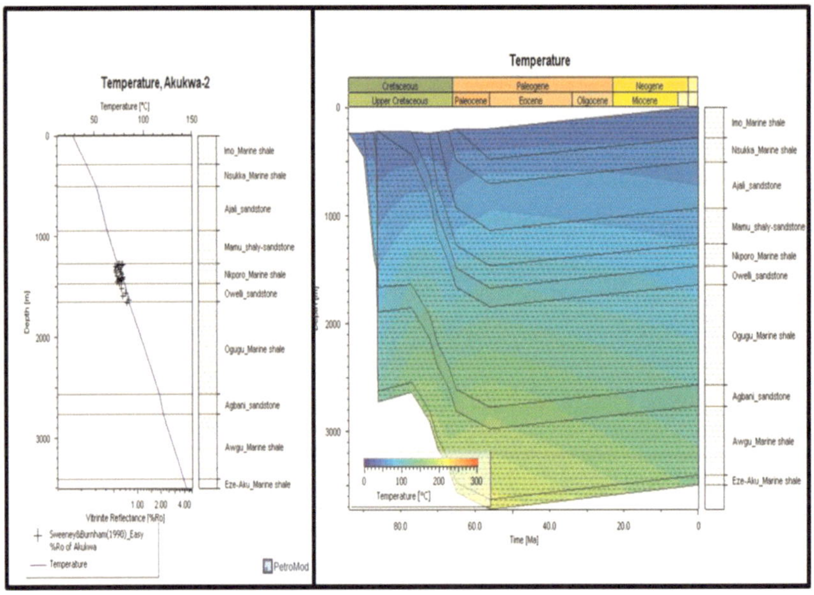

Figure 6.
Paleo-temperature modeling in Akukwa-2 well calibrated using borehole temperature; showing correlation among Burial history with temperature overlay, measured temperature and modeled Temperature for the studied well, notice that the maximum temperature values of 120–145°C in the area was attained between mid-Paleocene and mid Miocene (60-15ma) on Coniacian Agwu Source strata, higher temperatures are associated with Santonian tectonic episode.

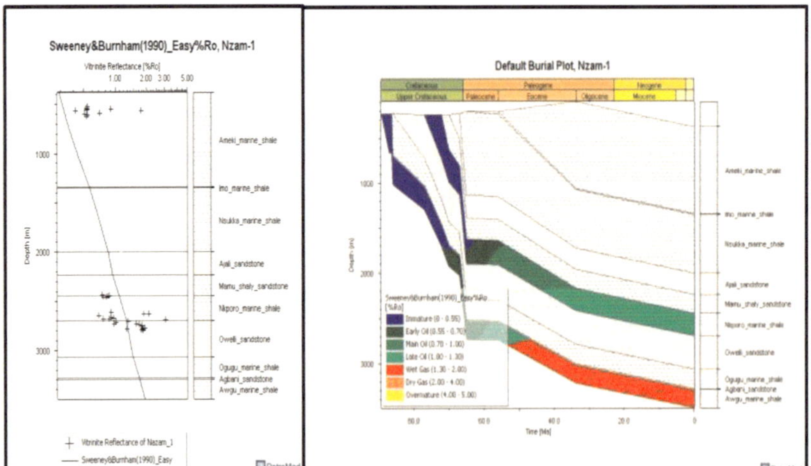

Figure 7.
Burial and thermal maturity histories of the Coniacian and upper Campanian source rocks for the studied well (Nzam-1 well) showing the positions of the oil window and the various hydrocarbon generation phases. Notice that Agwu Source rocks have entered the post maturity gas evolution and still generating gas to present day while the late Campanian Nkporo Source has entered the oil window and has marginal maturity. To the right the depth versus mean vitrinite reflectance plot indicating reasonable correlation between the measured and the modeled vitrinite reflectance.

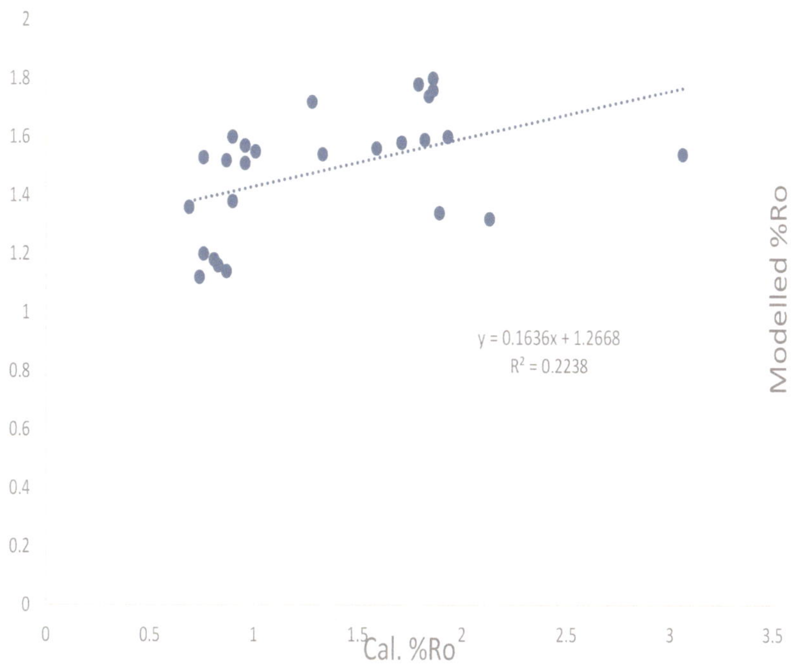

Figure 8.
Showing a reasonable correlation between the modeled and the calculated vitrinite reflectance and the R^2 of 0.2238 shows the correctness of the model.

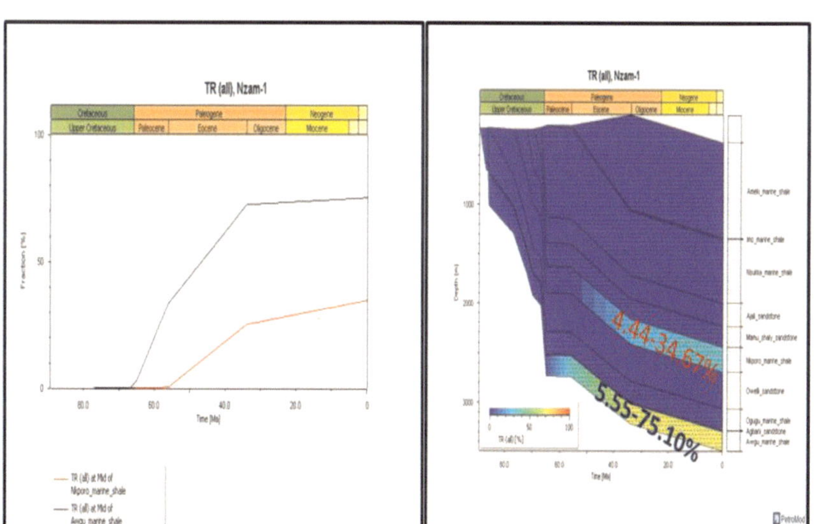

Figure 9.
Evolution of the transformation ratio and rate of hydrocarbon generation with age from the Coniacian and late Campanian source strata in the studied well (Nzam-1 well). Notice the Coniacian Agwu Shale has a higher transformation ratio which lies between 5.55 and 75.10% and greater than that Nkporo Source strata with transformation ratio between 4.44 and 34.67%.

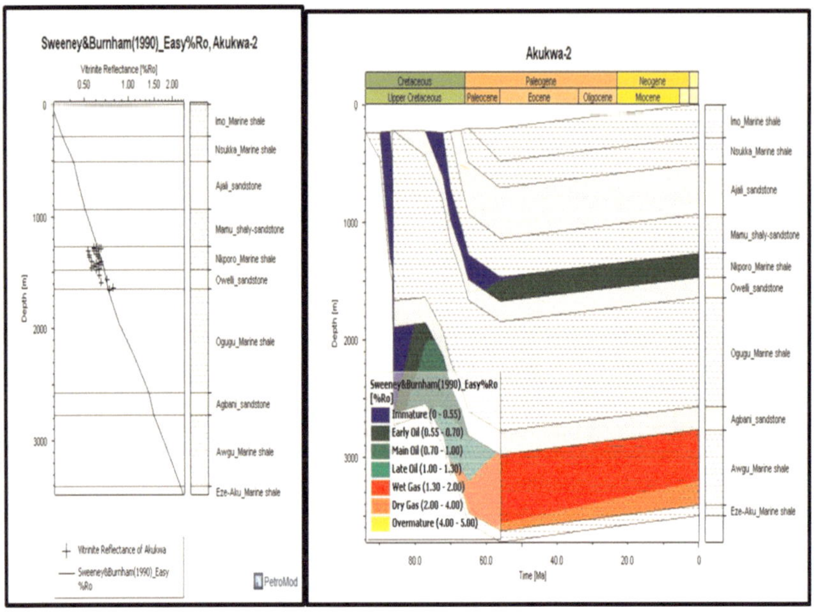

Figure 10.
Burial and thermal maturity histories of the Coniacian and late Campanian source rocks for the studied well (Akukwa-2 well) showing the positions of the oil window and the various hydrocarbon generation phases. Notice that Agwu Source rocks have entered the postmaturity gas evolution and still generating gas to present day while the late Campanian Nkporo Source has entered the oil window and is at the early generation stages. To the right the depth versus mean vitrinite reflectance plot indicating reasonable correlation between the measured and the modeled vitrinite reflectance.

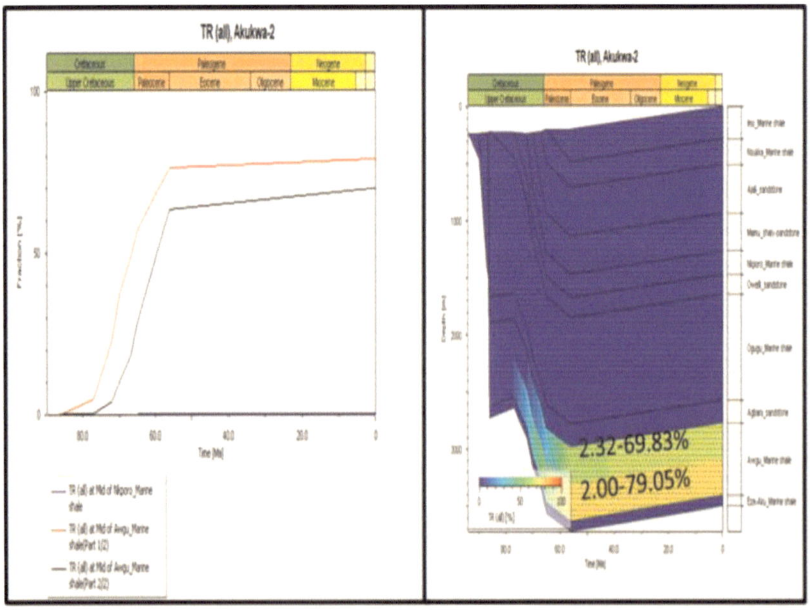

Figure 11.
Evolution of the transformation ratio and rate of hydrocarbon generation with age from the Coniacian-late Campania source rocks in the studied well (Akukwa-2).

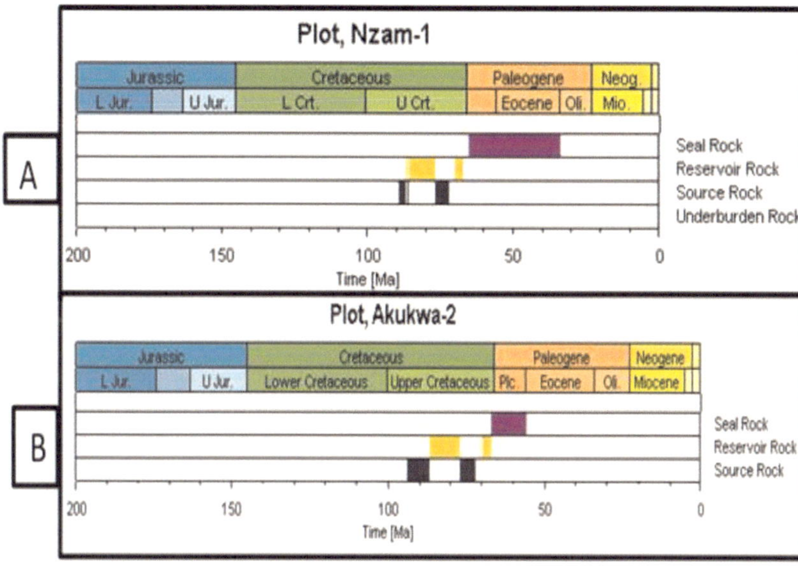

Figure 12.
Modeled Petroleum system elements of the studied (A) Nzam-1 and (B) Akukwa-2 Wells respectively; Both showing the positions of the two source strata under study (Coniacian Awgu and late Campanian Nkporo Source rocks, the positions of the Reservoir Rocks (Coniacian Agbani Sandstone and the Campanian Owelli Sandstone) and the positions of the Paleogene seals (majorly Imo Shale) all of which have demonstrated favorable conditions for Petroleum accumulations.

Figure 13.
Heavy crude seepage from the Owelli Sandstone at Egwueme, Lower Benue Trough [1].

and it gave way to the tectonic activities that took place in Santonian times, which resulted in the folding and uplifting of the Abakaliki Sector of the Trough and the subsidence of Anambra Platform. The latter event led to the formation of the

Anambra Basin and this constituted a major depocentre of clastic sediments and deltaic sequences [17, 18]

Based on the burial/thermal history model; the burial temperature within Agwu Formation in Nzam-1 well ranges from 30 to 145°C and that of Nkporo Formation ranges 28 to 125°C (**Figure 5**); From the burial/Thermal history of Akukwa-2 well the burial temperature in Agwu and Nkporo Formation ranges from 29.5 to 145°C and 28.5 to 95°C (**Figure 6**) respectively.

From the burial history model in Nzam-1 well (**Figure 5**), it was observed that before the Santonian tectonic episode (Pre-rift periods), the Kerogen in the Coniacian Awgu Shale experienced lower temperatures between the ranges of 30 and 55°C, it further experienced increased temperatures ranges between 120°C and 145°C between the Santonian and Miocene (83-15Ma) and (post-rift Periods) while in the burial model of the Lower Benue Sector in Akukwa-2 (**Figure 6**), the Kerogen in the Coniacian source strata has experienced lowest temperatures ranges of between 29.5 and 50°C before the Santonian times (Pre-rift) and exponential increased temperature ranges of 120 and 145°C between Santonian (83ma) up to present day. Consequently, the relatively high temperature and geothermal gradients experienced by the Coniacian Awgu strata between the Santonian and Miocene times aided the quickening of the organic matter maturation, oil generation and subsequent cracking of the oil to form gas. Since the temperature continued to decrease exponentially to the present day and such decrease in temperature have favored the preservation of the gas reservoirs and therefore the survival of hydrocarbons in the deeper strata can be guaranteed. More so the geothermics, the configuration of the hydrocarbon generation timing and reservoir cap development has favored accumulation (**Figure 12**) and the low geothermal field background after the formation of cracked gas has increased its chances of survival in their respective reservoirs within the Lower Benue Trough due to basin cooling and the resulting present day low heat flow (48m/Wm$^{-2)}$. The moderate temperature values of up to 125°C of Nkporo Formation for the studied Nzam-1 has aided its maturity and this happened due to basin heating and considerable burial depth which has placed the organic matter of Nkporo on an advantage position to mature relative to the organic matter of the same Formation in Akukwa-2 well with the highest temperature value of less than 95°C has entered the threshold of oil generation in Late Maastrichtian (65Ma) and continues to remained within the onset of generation owing to lack of sufficient heat and shallow burial depth that is needed for it to exit this present hydrocarbon generation phase.

It may therefore be said that Temperature evolution affected the development of organic matter pores in strata and their gas adsorption capacity. In sum, a higher temperature and a greater extent of thermal evolution should have resulted in highly developed organic matter pores [19], whereas a lower temperature should have enhanced the gas adsorption capacity. As shown in **Figures 5** and **6**, the high temperature of the Coniacian Agwu source rocks, which may be attributed to the high temperature gradient and deeper burying level of these rocks and more so predates Santonian tectonic episode, was highly favorable for the maturity of organic matter and the development of their pores, whereas the increased heat process after the main hydrocarbon aided the oil cracking gas generation periods but however the reduction in heat benefited the preservation of source rock gas should the cooling continues.

The one-dimensional burial and maturity model of Nzam-1 and Akukwa-2 wells was modeled after [14, 15] kinetic models to ascertain the hydrocarbon generation potential of upper Cretaceous organic rich shaly intervals.

Once source rocks have reached their expulsion threshold, they may expel the hydrocarbons upward into the reservoirs [19]. The timing of hydrocarbon generated and expelled from the Coniacian Awgu and upper Campanian Nkporo source rocks were modeled. Oil generation is defined in this study by transformation ratios between 10% and 50% [19]. Immature source rocks have transformation ratios less than10% (no generation). Peak oil generation occur at a transformation ratio of 50% when the main phase of oil generation is reached. And transformation ratio greater 60% when gas phase of the hydrocarbon generation is reached. The calculation of the transformation ratios is based on the Vandenbroucke et al., (1999)-TIII- (North Sea) kinetic model for the Coniacian Agwu and late Campanian Nkporo source rocks (**Figures 9** and **11**).

From the burial history model of Nzam-1 well with transformation overlay (**Figure 9**), it is clear that the Awgu has a higher transformation ratio as compared to Nkporo and this is consequent to their thermal maturity which is confirmed in the Transformation-Time plot with Awgu shale having higher curve. The transformation ratio of Awgu and Nkporo Shales in Nzam-1 well is 5.55–75.10% and 4.44–34.67% respectively (**Figure 9**) while the transformation ratios for Agwu Formation in Akukwa-2 well are 2.32–69.83% for upper strata and 2.00–79.05% for the lower strata while the Nkporo Formation had no transformation ratio (**Figure 11**). The transformation- Time plot shows that the thermal maturity and the transformation of organic matter increase with time and depth. Therefore, a reasonable correlation can be drawn between the burial plot and Transformation Time plot (**Figures 9** and **11**).

The hydrocarbon generation and expulsion modeling of Nzam-1 well model shows that the Agwu Shale reached early phase of oil generation in late Campanian and extended from 75 Ma to 69 Ma. Subsequently, main phase of oil generation began during early Paleocene and extended from 65 Ma to 62 Ma. The gas phase began at mid Eocene and extended from 48Ma to present day. The model has also shown that the expulsion of hydrocarbon from Agwu Source rocks occurred between 62 Ma up to present day and peak expulsion at 34 Ma during late Eocene (**Figures** 7 and **9**).

As for the late Campanian Nkporo Source rock, the Model of Nzam-1 has shown that the Nkporo Shale reached early phase of oil generation in early Paleocene and extended from 65 Ma to 60 Ma. Subsequently, main phase of oil generation began during late Paleocene and extended from 56 Ma to 42 Ma. The model has also shown that the expulsion of oil from Nkporo Source rocks occurred between 42 Ma to present day and peak expulsion at 30 Ma during early Oligocene (**Figures** 7 and **9**).

The hydrocarbon generation and expulsion modeling of Akukwa-2 well shows that the Agwu Shale reached early phase of oil generation in early Santonian and extended from 85 Ma to 78 Ma. Subsequently, main phase of oil generation began during late Campanian and extended from 75 Ma to 70 Ma. The gas phase began at mid Paleocene and extended from 58Ma up to present day. The model has also shown that the expulsion of hydrocarbon from Agwu Source rocks occurred between 70Ma and up to present day and peak expulsion at 57 Ma during late Paleocene (**Figures 10** and **11**).

The hydrocarbon generation and expulsion model of Akukwa-2 well has also shown that the late Campanian Nkporo Source rocks has just entered the early phase of oil generation late Maastrichtian (67Ma) to present day; However, the late Campanian Nkporo Formation did not reach the main phase of oil generation and expulsion owing to the fact that it lacks the requisite burial depth, temperature and pressure in favor of oil generation and expulsion prevent further generation and expulsion in the upper Campanian source strata in this Formation (**Figures 10** and **11**). From the above, it can be said that the organic matter in Coniacian Awgu source rocks has reached the post maturity evolution phase and the expected hydrocarbon product is gas. While the late Campanian Nkporo source rocks on the other hand is at early maturity to peak maturity evolution Phase and the expected hydrocarbon product is oil.

Petroleum accumulation depends on the configuration of hydrocarbon generation as well as migration, Formation, and evolution of the reservoir and its sealing conditions [19]. Considering the Petroleum System Elements (PSE) models, the Coniacian Awgu source rocks reached their oil and gas generation peaks in the Late Paleocene to early Eocene (58-52ma) and continued to generate gas to present day in the Nzam-1 model (**Figures 7, 9**, and **12**) whereas reached its peak generation in late Campanian to msid Paleocene (80-58ma) and continued to generate gas to present day in Akukwa-2 well (**Figures 10–12**) and the generation took place later than the formation of the regional reservoir rocks (Coniacian Agbani and late Campanian Owelli Sandstone Members) and the Paleogene seal rocks majorly Imo Shales (**Figure 10**). While the late Campanian Nkporo source rocks reached their oil generation peaks in the early Eocene to mid Oligocene (55-32ma) and continued to present day in Nzam-1 well (**Figures 7, 9** and **12**) and only entered the hydrocarbon generation threshold up to present day in Akukwa-2 well (**Figures 10–12**). The Petroleum System Elements (PSE) model revealed a viable petroleum system comprising of two source rocks dated Coniacian (89Ma) and Campanian (78Ma); two reservoir rocks dated also Coniacian (88Ma) and Campanian (78Ma) and the cap rock is dated Paleocene (65Ma) which indicates that hydrocarbon generation and expulsion occurred later than the formation of the regional reservoir rocks basically Coniacian Agbani and late Campanian Owelli Sandstone members and the seal rocks majorly the Juxtaposed Paleocene Imo Shales. The hydrocarbon generation period provided favorable conditions for the accumulation of oil and gas from these Coniacian and late Campanian source rocks in the Lower Benue Trough. Given the multiple phases of hydrocarbon generation, the Coniacian Agwu source rocks had more favorable conditions for hydrocarbon accumulation than those of the late Campanian Nkporo Source rocks (**Figure 12**). The Paleogene Shale Units within the Lower Benue Trough are the most important oil and gas seal. Therefore, it is possible hydrocarbons were generated and expelled from Agwu Shale and evidence can be seen in heavy crude seepage from the Owelli Sandstone at Egwueme, Lower Benue Trough. And valid petroleum system elements exist in the basin with relative differences in the generation and expulsion periods of hydrocarbon generation.

A plot of measured vitrinite reflectance against the modeled vitrinite reflectance was computed and the RSME is found to be 0.2238 which indicates good correlation and hence the correctness of the model (**Table 6** and **Figure 8**).

From the Plot of Hydrogen Index (HI) against Maximum Temperature (Tmax), calculated Vitrinite Reflectance after [11] and calculated Vitrinite Reflectance after [16] showing Kerogen quality and thermal maturity stages for the studied sediments Nkporo Marine Shale and Agwu Marine Shales in the Lower Benue Trough sedimentary section of the Nzam-1 well indicated that Kerogen from Nkporo is immature to early maturity and is also Predominantly type II-III kerogen which is mixed oil/gas prone with minor occurrences of type II and type III While kerogen from Agwu shale has attained peak maturity to post maturity and the organic matter is predominantly consist of type II-III kerogen perhaps with minor occurrences of type II Kerogen (**Figure 14**).

Cross plot of Production Index against Tmax revealed that samples from Agwu Formation fall within the oil and gas window and the organic matter has experienced high level conversion which is also an indication of Peak maturity to post maturity (**Figure 15**).

Measured %Ro	Modeled %Ro	Model	Residual	Residual Square	RMSE
0.74	1.12	1.387864	0.267864	0.071751122	0.178092
0.87	1.14	1.409132	0.269132	0.072432033	
0.83	1.16	1.402588	0.242588	0.058848938	
0.81	1.18	1.399316	0.219316	0.048099508	
0.76	1.2	1.391136	0.191136	0.03653297	
0.9	1.6	1.41404	−0.18596	0.034581122	
2.13	1.32	1.615268	0.295268	0.087183192	
1.89	1.34	1.576004	0.236004	0.055697888	
0.69	1.36	1.379684	0.019684	0.00038746	
0.9	1.38	1.41404	0.03404	0.001158722	
0.96	1.51	1.423856	−0.08614	0.007420789	
0.87	1.52	1.409132	−0.11087	0.012291713	
3.06	1.54	1.767416	0.227416	0.051718037	
0.76	1.53	1.391136	−0.13886	0.01928321	
1.33	1.54	1.484388	−0.05561	0.003092695	
1.01	1.55	1.432036	−0.11796	0.013915505	
1.59	1.56	1.526924	−0.03308	0.001094022	
0.96	1.57	1.423856	−0.14614	0.021358069	
1.71	1.58	1.546556	−0.03344	0.001118501	
1.82	1.59	1.564552	−0.02545	0.000647601	
1.93	1.6	1.582548	−0.01745	0.000304572	
1.28	1.72	1.476208	−0.24379	0.059434539	
1.84	1.74	1.567824	−0.17218	0.029644575	
1.86	1.76	1.571096	−0.1889	0.035684721	
1.79	1.78	1.559644	−0.22036	0.048556767	
1.86	1.8	1.571096	−0.2289	0.052397041	

Table 6.
Showing the calculated and the modeled vitrinite reflectance values with the resultant Root mean square error of 0.178092

And from the cross plot of Production Index against Tmax and calculated Vitrinite Reflectance after [11] and [16] suggested that the Kerogen from Agwu Formation has undergone an intensive generation expulsion and the samples fall within the oil window and gas window, it also indicated that the organic matter has undergone high level conversion and has entered the over mature zone (**Figure 15** and **16**). This implies that the expected hydrocarbon type is oil, Condensate-wet gas and dry gas. It can therefore be inferred that the samples from Agwu Formation of the Nzam-1 Well have attained peak to post thermal maturity.

The measured Vitrinite reflectance (Ro) in both Nzam-1 and Akukwa-2 wells (**Figures 7, 8** and **10**) has a reasonable correlation with the modeled vitrinite reflectance after [14]. The heat flow histories used in the calculations are also plotted in

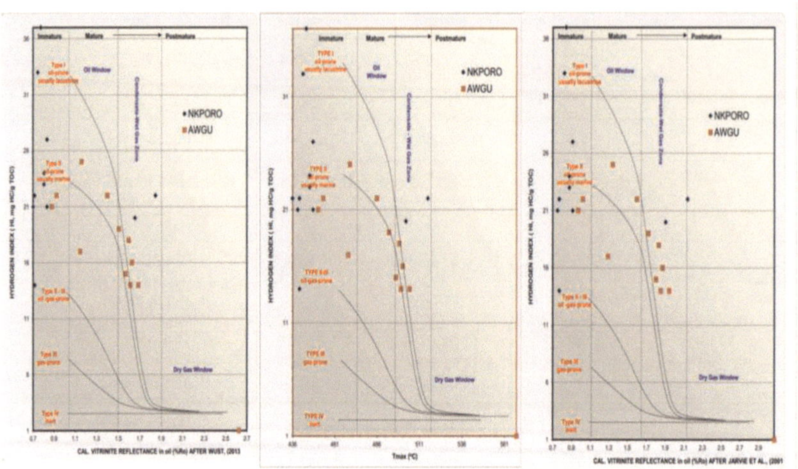

Figure 14.
Plot of Hydrogen Index (HI) against Maximum Temperature (Tmax), calculated Vitrinite Reflectance after [11] and calculated Vitrinite Reflectance after [16] showing Kerogen quality and thermal maturity stages for the studied sediments Nkporo Marine Shale and Agwu Marine Shales in the Lower Benue Trough sedimentary section of the Nzam-1 well indicating that Kerogen from Nkporo is immature to early maturity and is also Predominantly type II-III kerogen which is mixed oil/gas prone with minor occurrences of type II and type III While kerogen from Agwu shale has attained peak maturity to post maturity and the organic matter is predominantly consist of type II-III kerogen perhaps with minor occurrences of type II Kerogen.

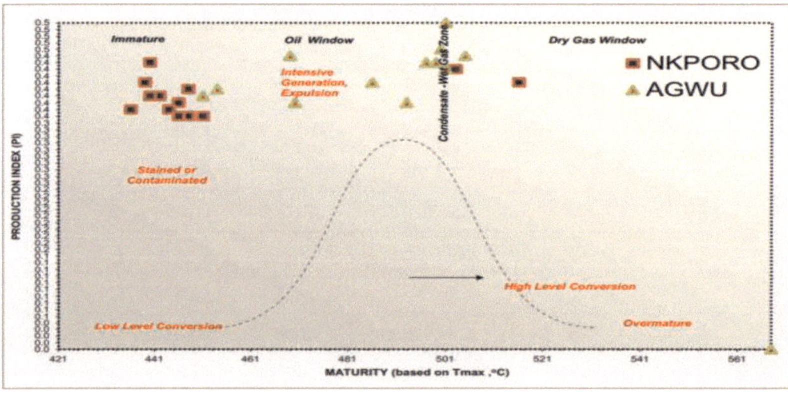

Figure 15.
Plot of Production Index against Tmax showing Kerogen Conversion/Maturity of late Campanian Nkporo Shales at immature-early mature stages and Coniacian Awgu Shales at peak maturity to post maturity stages in the sedimentary section of the Lower Benue Trough, Nzam-1 well indicating that the Nkporo Shale is at early maturity stages while Awgu is at peak/post maturity stages and has expelled hydrocarbon.

(**Figures** 2 and 4). In Nzam-1 well, heat flow values range between 48 and 72 mW/m^2 while in Akukwa-2 well it ranges 48 and 75 mW/m^2 and this can be attributed to the variability of heat flow and geothermal gradient in the earth subsurface.

6. Conclusions

Results of the investigation of burial/thermal histories and timing of petroleum generation/expulsion modeling of the Coniacian Agwu and the late Campanian

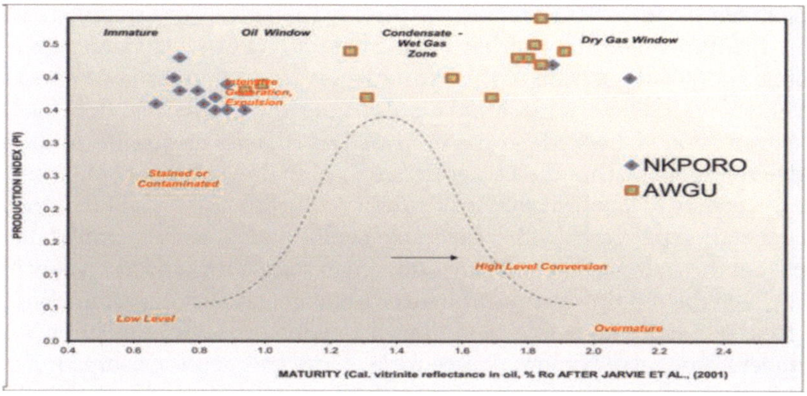

Figure 16.
Plot of Production Index against calculated Vitrinite reflectance after Jarvie at al., 2001 showing Kerogen Conversion/Maturity of late Campanian Nkporo Shales at immaturity-early maturity stages and Coniacian Awgu Shales at peak to post maturity stages in the sedimentary section of the Lower Benue Trough, Nzam-1 well indicating that the Nkporo Shale is at immaturity-early maturity stages while Awgu is at peak/post maturity stages and has expelled hydrocarbon.

Nkporo source rock intervals using Schlumberger's one dimensional modeling software indicates that:

The maximum temperature has been recorded in Agwu Formation where it reached 145°C at depth of 3400m and is responsible for the cracking of the oil to gas. There is also a strong indication that the high temperatures within the Formation were occasioned by deeper burial depth and possibly the Santonian tectonic episode which increased the geothermal gradient in the area. These further confirmed that the end of the Cretaceous era saw uplift, denudation and subsidence in the Lower Benue Trough as well as increased geothermal gradient to have caused complex processes of uplift, denudation, basin heating and cooling.

The hydrocarbon generation and expulsion modeling of the Lower Benue Trough shows that the Agwu Shale reached early phase of oil generation in late Campanian and extended from 75 Ma to 69 Ma in Nzam-1 well while it occurred at early Santonian and extended from 85 Ma to 78 Ma in Akukwa-2 well., Subsequently, main phase of oil generation began during early Paleocene and extended from 65 Ma to 62 Ma in Nzam-1 well and began during early Paleocene and extended from 75 Ma to 70 Ma in Akukwa-2 well. The gas phase began at mid Eocene and extended from 48Ma to present day in Nzam-1 well and began at mid Paleocene and extended from 58Ma to present day in Akukwa-2 well. As for the late Campanian Nkporo Source rock, the Model of Nzam-1 well has shown that the Nkporo Shale reached early phase of oil generation in early Paleocene and extended from 65 Ma to 60 Ma. Subsequently, main phase of oil generation began during late Paleocene and extended from 56 Ma to 42 Ma, whereas hydrocarbon generation and expulsion model of Akukwa-2 well has also shown that the late Campanian Nkporo Source rocks has entered the early phase of oil generation in late Maastrichtian (67Ma) to present day and did not reached the main phase of oil generation and expulsion owing to the fact that it lacks the requisite burial depth, temperature and pressure in favor of oil generation and expulsion.

The above shows that Awgu source rocks are overmature and are at post maturity gas evolution stage, generating gas to the present day whereas the Nkporo Shales are at early to peak maturity stage generating and expelling oil to the present day. Since the temperature continued to decrease exponentially to the present day and such decrease in temperature have favored the preservation of the gas reservoirs and therefore the survival of hydrocarbons in the deep strata can be guaranteed. More so the geothermics, configuration of the hydrocarbon generation timing and reservoir cap development has favored accumulation and the low geothermal field background after the formation of cracked gas has increased its chances of survival in their respective reservoirs within the Lower Benue Trough due to basin cooling and the resulting present day low heat flow (48 $m/Wm^{-2)}$. It can also be said that the hydrocarbon generation period provided favorable conditions for the accumulation of oil and gas from these Coniacian and late Campanian source rocks in the Lower Benue Trough. Given the multiple phases of hydrocarbon generation, the Coniacian Agwu source rocks had more favorable conditions for hydrocarbon accumulation than those of the upper Campanian Nkporo Source rocks. Drift sediments accompanying Coniacian- late Campanian periods possibly formed the major reservoir for the expelled hydrocarbons of the Agwu and Nkporo Shales and could represent potential reservoir units (Agbani and Owelli sandstones member) for hydrocarbons generated in the upper Cretaceous and time equivalent source rocks of Lower Benue Trough. This study suggest that valid petroleum systems exist in the basin with relative differences in the generation and expulsions periods of hydrocarbon generation in the two Formations and Paleogene Shale unit especially Imo Shales is the most important oil and gas cap rocks sealing the Owelli and Agbani sandstones that are being charged by the Coniacian Agwu and the late Campanian Nkporo Shales.

It is recommended that exploration processes be focused on the known deeper location of the Agwu and Nkporo Formations to predict the source kitchen. A 2D and 3D Burial History and Maturity modeling of the upper Cretaceous sediments of the Lower Benue Trough should be carried out and interpreted so as to identify the possible migratory path ways of the generated hydrocarbon by integrating structural information and hydrocarbon timing and generation. This will require the integration of Seismic, Geochemical and stratigraphic information.

Author contributions

Conceptualization and Data curation: Danwazan, J. and Adeleye, M. A.
Writing-original draft: Danwazan, J.
Writing-review & editing: Danwazan J. and Adeleye M.A,
Resources: Adeleye, M. A.
Investigation: Danwazan, J.
Supervision: Adeleye, M. A.

Author details

Jerry Danwazan[1,2]*, Adesina Mutiu Adeleye[2], Trust-Paul Bakyu[2], Mamidak M. Iliya[1] and Joseph Kehinde Egunjobi[2]

1 Nasarawa State University, Keffi, Nigeria

2 University of Ibadan, Ibadan, Nigeria

*Address all correspondence to: danwazanjerry1000@gmail.com

References

[1] Abubakar MB. Petroleum potentials of the Nigerian Benue Trough and Anambra Basin: Regional studies. Natural Resources. 2014;**5**:25-58

[2] Tissot BP, Welte DH. Petroleum Formation and occurrence. 2nd ed. Berlin: Springer-Verlag; 1984. p. 699

[3] Ehinola OA, Ajala AA. Thermal Maturity and burial history modelling of Anambra basin. In: 6th NAPE-UAP Mini-conference Book of abstract A7. Southeastern Nigeria; Nigerian Association of Petroleum Explorationists (NAPE); 2012

[4] Oluwajana OA, Ehinola OA. Hydrocarbon-charge modelling of Anambra basin, Southeastern, Nigeria: Implications for Cretaceous-sourced plays. Arabian Journal of Geoscience. 2016;**9**(3):1-16

[5] Akaegbobi MI, Adegoke KA, Onyehara TI, Adeleye AM. Organic geochemical characterisation of the Campano-Maastrichtian sediments in Anambra Basin, SE, Nigeria: Implication for Paleodepositional Conditions, Provenance and Petroleum Generation Potential. Journal of Environment and Earth Science. 2017;**7**:18

[6] Adeleye MA, Adeniyi VA, Oke AO. Aspects of hydrocarbon potential of the Tertiary Imo Shale Formation in Anambra Basin, Southeastern Nigeria. A Journal of Applied Geology and Geophysics. **2017**;5:74-83

[7] Adeleye AM, Yuhong L, Abiodun AJ. Geochemical Evaluation of Nkporo Formation from Nzam-1 Well, lower Benue Trough. Journal of Petroleum and Coal. 2016;**58**(3):328-338

[8] Adebayo FO, Adeleye AM, Adegoke KA. Paleonenvironmental Reconstruction and Hydrocarbon Potentials of Upper Cretaceous Sediments in the Anambra Basin, Southeastern Nigeria. International Journal of Coal Geology. 2018;**192**:56-72

[9] Lopatin NV. Temperature and geologic time as factors in coalification. Altai Izvestiya AkademiyaNauk Kazakh SSR. SeriyaGeologicheskaya. 1997;**3**:95-106

[10] Littke R, Buker C, Luckge A, Sachsenhofer RF, Welte DH. A new evaluation of palaeoheat flows and eroded thicknesses for the Carboniferous Ruhr Basin, western Germany. International Journal of Coal Geology. 1994;**26**:155-183

[11] Jarvie DM, Claxton BL, Henk F, Breyer JT. Oil and shale gas from the Barnett Shale, Fort Worth basin, Texas. AAPG Bulletin. 2001;**85**:A100

[12] Adeleye AM, Adeigbe OE, Yuhong L. Source Rock Evaluation of the Agwu Formation from Nzam-1 well, lower Benue Trough, Nigeria. NAPE Bulletin. 2014;**26**(1):87

[13] Hardenbol J, Jacquin T, Farley MB, de Graciansky P-C, Vail PR. Mesozoic and Cenozoic chronostratigraphy and cycles of sealevel changes. In: de Graciansky P-C, Hardenbol J, Jacquin T, Vail PR, editors. Mesozoic and Cenozoic Sequence Stratigraphy of European Basins. Vol. 60. SEPM Special Publication; 1998. pp. 3-13

[14] Sweeney JJ, Burnham AK. Evaluation of a simple model of vitrinite reflectance based on chemical kinetics. AAPG Bulletin. 1990;**74**(10):1559-1570

[15] Vandenbroucke M, Behar F, Rudkiewicz JL. Kinectic modelling of

petroleum formation and cracking: Implications from the high pressure/high temperature Elgin Field (UK, North Sea). Organic Geochemistry. 1999;**30**(9): 1105-1125

[16] Wust RAJ, Hackley PC, Nassichuk BR, Willment N, Brezovski R. Vitrinite Reflectance Versus Pyrolysis Tmax Data: Assessing Thermal Maturity in Shale Plays with Special Reference to the Duvernay Shale Play of the WesternCanadian Sedimentary Basin. Alberta, Canada: SPE Unconventional Resources Conference and Exhibition-Asia Pacific held in Brisbane, Australia; 2013

[17] Obaje NG. Geology and Mineral Resources of Nigeria. Berlin Heidelberg: Springer-Verlag; 2009. pp. 57-65

[18] Fatoye FB, Gideon YB. Geology and mineral resources of the lower Benue Trough Nigeria. Advances in Applied Science Research. 2013;**4**:21-28

[19] Al-Areeq NM, Albaroot MA. Source rocks evaluation and thermal maturity evolution of the dhamarali field, sab'atayn basin, Yemen. Asian Journal of Science and Technology. 2019;**10**(01): 9364-9374

Chapter 3

Possible Oil Spills Disposal for Environmental Water-Body Protection

Veronika Veľková, Helena Hybská and Tatiana Bubeníková

Abstract

The possibilities of the oil spill cleanup from the water environment are presented. Mechanical methods of oil recovery are described—the oil containment booms, oil skimmers, and use of sorbents. The sorption capacity of various sorbents is compared based on laboratory tests according to the ASTM F726 methodology. The results of the determination of residual oil pollution of water after the cleaning process are presented. The properties of the absorption sock during the sorption of crude oil and the oil/water mixture were also presented.

Keywords: oil spill, oil contamination, sorbents, absorbent sock, sorption capacity

1. Introduction

Water and soil represent important components of the environment, which are strongly affected by contamination with foreign substances. Oil and oil products are important contaminants. Oil substances get into water and soil in various ways, either during the normal use and processing of oil or during accidental spill. Oil contamination is known mainly "thanks" to large-scale oil spills in oil tanker accidents (Exxon Valdez Accident in 1987) or oil platforms (Deepwater Horizon Blowout in 2010), where after the release of several million tons of oil, large and long-term damage to the environment occurred at sea and on the coast, such as also to damage the economy in the given areas [1–3]. There are some of the greatest oil spills in the history in the **Table 1**. The greatest accidental oil spill occurred in the 2010 after blow out at the BP's oil rig Deepwater Horizon. More than 200 million gallons of oil spilled into the Gulf of Mexico, which meant enormous damage to the aquatic ecosystems, the coastal ecosystems, the life of aquatic and terrestrial organisms, and the life of the entire society in Louisiana. The fishing in the Gulf was damaged, the food industry collapsed, and also the tourism industry completely fell. The economic and social life of Louisiana was so destroyed, and the solution to the accident required such interventions that a special web page was set up to inform people [6, 7].

Despite the fact that the number of such large oil spills is decreasing as a result of the increase in safety during the transportation of oil and oil products, the improvement of technological processes, emergency situations, and gradual releases of oil

Incident	Location	Date	Oil amount
Iraq-Kuwait war	Persian Gulf	January 1991	380–520 mil. Gallons
BP Deepwater Horizon accident	Gulf of Mexico	April 22, 2010	206 mil. Gallons
Ixtoc I accident	Bay of Campeche off Ciudad del Carmen	June 3, 1979	140 mil. Gallons
Atlantic Empress accident	The seaside of Trinidad and Tobago	July 19, 1979	90 mil gallons
Castillo de Bellver accident	Saldanha Bay South Africa	August 6, 1983	79 mil. Gallons
ABT Summer accident	Off the coast of Angola	May 28, 1991	79 mil. Gallons
Amoco Cadiz accident	Portsall, France	March 16, 1978	69 mil. Gallons

Table 1.
The greatest oil spills in the history [4, 5].

substances into the environment occur continuously, so there is always room for the development of methods and techniques for removal of oil contamination [1]. Depending on the amount of leaked substances (the size of the oil spill), the environmental conditions (type of water source—calm or flowing water), and technical possibilities, the main cleaning procedures include oil booms, separators, and separators (skimmers), various types of sorbents, in-situ burning, chemical cleaning (dispersion of petroleum substances), and bioremediation [2].

2. Effects of oil spills to the water environment

Oil and petroleum substances represent the basic raw material for the needs of industry, and their impact on all components of the environment is not negligible [8]. The consequences of environmental pollution with oil substances can appear immediately or after a longer period of time.

An oil spill usually affects the entire affected region, destroys the aquatic environment and life in it, and if it reaches the coast, great damage appears there as well (**Figure 1**).

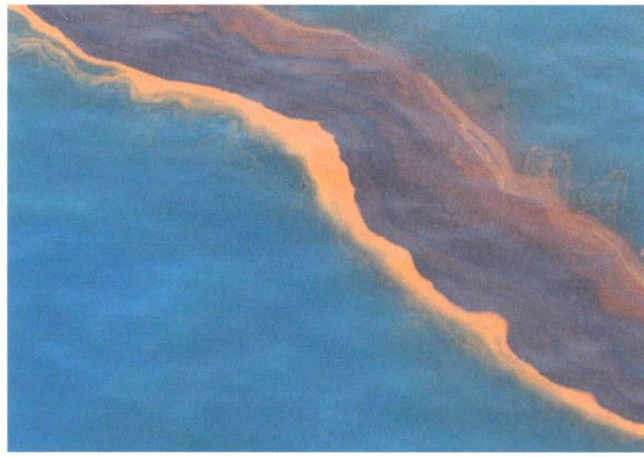

Figure 1.
Deepwater Horizon oil spill on May 12, 2010 [9].

Figure 2.
Oiled bird (© Guardian Unlimited) and oiled seal (© Tom Loughlin, NOAA) [10].

Oil spills directly or indirectly kill fish and aquatic organisms, birds, plants, affect the oxygen regime, disrupt the natural cycle of ecosystems, and change the physical and chemical properties of the aquatic environment (**Figure 2**).

2.1 Mechanism of oil spill pollution in water bodies

When oil leaks into an open water body, it forms a thin layer on the water surface. In the case of strong winds and strong currents, the oil layer can quickly cover a relatively large area. For example, 1 ton of oil can cover an area of up to 12 km^2 with a 1 mm thick layer [11]. The created layer is affected by various factors such as wind, waves, water currents, UV light, the presence of various types of microorganisms, and the oil weathering process occurs [11–13]. Weathering is a process in which leaked oil is gradually lost from the formatted spill depending on the current environmental conditions. The weathering process can be understood as a combination of various physical and chemical changes such as evaporation, dispersion, emulsification, sedimentation, oil entrainment to the water column, biodegradation [11–13]. The study in the year 1992 in the Exxon Valdez accident area shows that 20% of oil evaporated, 50% biodegraded, 14% was cleaned up, 13% remained in subtidal sediments, 2% remained on shorelines, and less than 1% remained in the water [1, 14].

The speed of the weathering process significantly depends on the strength and cohesion of the oil film on the surface. Yan et al. [12] studied the forces affecting the entrainment ability of petroleum substances into the water column. According to them, the difference between the surface tension of water and oil and the viscosity of oil are decisive. Waves on the water surface act on the oil layer, which "break" it. A single strong wave can work, but weaker repeated waves are also effective. A "sandwiching" of water and oil occurs, the oil film becomes thinner until it disintegrates [12, 13].

Oil pollution of seawater also affects the formation and properties of marine snow [15]. After the Deepwater Horizon accident (2010), a high formation of marine snow was observed in the Gulf of Mexico. Passow et al. [15] assume three possible ways of formation of marine snow: increased production of mucous webs by oil-degrading bacteria combined with floating oil particles on the surface; coagulation of oil particles produced by the interaction of oil with sedimenting particles; coagulation of phytoplankton with oil droplets into larger aggregates. They also pointed out that oil significantly affects the sinking and breaking up of marine snow, while environmental conditions also interact. Snow in the natural environment of the Gulf of Mexico, composed mostly of oil particles and low-density mucus, disappeared in about a month after its appearance, while under laboratory storage conditions in the dark and at about 4°C, it remained on the surface for up to 2 years [15].

2.2 Oil substances in the water environment

The change in physical and chemical properties is significantly influenced by free oil substances that create an oil film on the surface, which reduces the transfer of oxygen to the water. Another process in which oxygen is pumped out of the water is the ongoing microbial oxidation of oil pollution. The result of both processes is a change in the concentration of dissolved oxygen, which causes a change in the course of chemical reactions in the process of photosynthesis, which is very unfavorable for the life of aquatic organisms, algae and other plankton are strongly affected [16].

Chemical composition of oil is very variable, present are straight and branched alkanes, alkenes, aromatic compounds, oxygen, and sulfur compounds and in small amount also nitrogen compounds and some metals. Polycyclic aromatic hydrocarbons have a significant presence, which have toxic, carcinogenic, and mutagenic effects. The easier ones remain in the oil film on the surface or are dispersed in the water column, the more difficult to settle in the sediments, and represent long-term hidden dangers. It is important to know the localization and persistence of these substances in nature [16, 17]. When analyzing the composition, it is possible to focus on some specific substances or to monitor the entire group according to normative methods, for example, PAHs or nonpolar extractable substances (NES) [16].

The acceptable concentration limit of these compounds can be determined using bioassays on appropriately selected sensitive organisms. Ecotoxicological studies have shown the impact of harmful effects of chemical substances on living organisms [18] and are one of the main tools suitable for assessing the effects of specific chemical compounds on environmental components. For acute and chronic toxicity tests, test organisms such as fish, daphnia, rats, birds, and seeds are suitable choices [19]. Due to their low price and good sensitivity (germination), the seeds of suitable plants are important toxicological tests for assessing the effects of toxic substances or organic inhibitors [20]. Tamada [21] monitored biodegradability by respirometric tests and compared different levels of toxicity of lubricating oils using toxicological tests. Tests were performed using earthworms (*Eisenia andrei*), arugula seed (*Eruca sativa*), and lettuce seed (*Lactuca sativa*) in mineral, synthetic, and used lubricating oil for different periods of their biodegradation in soil. Toxicity tests were used to indirectly measure the biodegradation of pollutants. The used lubricating oil proved to be the most toxic. Mineral and synthetic oils were efficiently metabolized in the soil, although they were still toxic after 180 days [21]. In their work, Cecutti and Agius [22] present the results of a study in which they successfully applied test organisms such as algae, pearl oysters, and fish to assess the ecotoxicological properties of various oils, including bio-oils new—before use and used—after 1000 hours of use in the aquatic environment. Bordulina [23] report the results of laboratory experiments, where they studied the effect of different concentrations of oil in the water environment on the algae *Chlorella vulgaris* Beijer and the abalone *Daphnia magna* Straus in model samples. They found that as the concentration of the oil increased, the survival of the test organisms decreased. Martinez [24] found acute toxic damage to daphnia after a 48-hour test. Authors, Bordulina [23] and Hybská et al. [25], confirmed that as the concentration of oil in the water increases, the survival rate of abalone decreases. They found that at a concentration of 0.2 mg of oil/1 l of water, regardless of its origin (mineral and bio oil—sunflower oil), there was no difference in the number of immobilized individuals [24, 26].

3. Oil spills cleanup techniques

A series of procedures and methods can be used to stop oil spills and their subsequent removal from the environment. These are mechanical methods for stopping the spread and collection of contaminating substances, chemical methods for the purpose of dispersion and emulsification of oily substances, burning in-situ, and bioremediation using microorganisms. Mechanical containment with the subsequent recovery of oily substances is one of the most used procedures with relatively high efficiency both in the USA and in European countries [14, 27].

3.1 Oil containment booms

The main function of the oil spill containment boom is "to close" the oil spill and to prevent from further spreading and floating oil to larger surface. It helps protect coasts or precious ecosystems from contamination. An oil containment boom represents a physical barrier that will ensure the containment and concentration of floating oil contaminants to enable their collection. It is usually the first resource to be deployed and the last to be removed when solving an accident [1, 14].

Oil containment booms come in many shapes, sizes, and lengths from smaller and lighter designed for manual use to deal with smaller leaks to robust, huge ones designed for use on the open sea using ships and ship cranes (**Figures 3** and **4**). In general, they all consist of the following basic parts [14].

- freeboard floating above the water surface contains the oil and prevents from slash-over

Figure 3.
Use of the oil containment boom [28].

Figure 4.
The oil containment boom in a ship lock [14].

- subsurface skirt below the surface

- rope or chain to stabilize the boom

The effectiveness of containment booms is significantly determined by the movement of water on the stream and the surrounding conditions (currents, waves, wind). The boom should be flexible enough to resist water movement and retain so much oil as possible [3, 14].

3.2 Oil skimmers

The next step in cleanup process after containing of oil spill is recovery of oil substances as much as possible from the water environment. Oil skimmers are intended for collecting oil from the water surface. Oil skimmers are devices floating on the surface and collecting oil substances [14, 29]. They consist of two parts—the device that keeps the oil skimmer on the surface and the collecting part itself. Depending on the conditions, especially the roughness of the surrounding water, self-propelled skimmers can be used, skimmers controlled from the shore or from the vessel. Based on the construction, we are talking about [29]:

- weir skimmers (**Figure 5**)—they remove oil from the water surface via a weir. The disadvantage is that they collect a relatively large amount of water at the same time, and they also collect water even when the oil is no longer present.

- drum or disc skimmers (**Figure 6**)—they collect oil from the surface on the principle of adhesion to metal or plastic disks or oleophilic belts or ropes.

The collected oil is usually stored in a containment tank or an additional temporary storage device is required. In some cases, it is necessary to heat the tanks in order to preserve the good ability of the collected oil to flow. The effectiveness of the oil recovery via a skimmer is determined by how quickly the skimmer can collect the oil and how well it minimizes the water–oil ratio [14]. The operation of skimmers can

Figure 5.
Weir oil skimmer [30].

Figure 6.
Drum oil skimmer [31].

significantly limit the occurrence of floating foreign material on the surface, e.g., branches, leaves, algae, grass, waste.

3.3 Sorbents

Another method of oil and oil products recovery is represented by sorbents. Sorbents are used to clean smaller oil spills, for final removing of oil substances or in places where skimmers cannot be used. Sorbents are solid, in water-insoluble material, which can absorb gaseous or liquid substances due to their physical and chemical composition. They work through the adsorption or absorption mechanism. Adsorbents become coated with oil substances on the surface, in pores and capillaries, binding can be both physical and chemical in nature. In order to be effective in oil spills disposal, sorption materials must be partly oleophilic and at the same time water-repellent. Absorbents pick up and retain liquids distributed throughout its molecular structure, typically is the volume increasing [2, 14, 32].

Sorbent material can be divided in three basic groups: natural organic material, natural inorganic, and synthetic.

Natural organic sorbents are mostly carbon-based products, such as straw, sawdust, coconut fibers, peat moss, etc., which can be in granular or fibrous form. These materials can adsorb more three times their weight of oil. Some of organic sorbents tend to break down to smaller particles, and it could be complicated to collect it. Other organic materials could sink to the bottom due to water absorption [14, 32].

Natural inorganic sorbents are used as granular materials such as perlite, clay, sand, ash. They can adsorb more than four times their weight in oil. As with organic sorbents, these materials also can have collection problems due to the breakdown into smaller particles.

Synthetic sorbents are composed of man-made materials, especially plastics such as polyurethane, polypropylene, polyethylene modified to adsorb liquids like a sponge. Other synthetic sorbents include cross-linked polymers and rubber materials, which absorb liquids into its solid structure, causing the sorbent material to swell [14]. Synthetic sorbents can absorb up 40–50 times their own weight in oil [14, 33].

3.3.1 Oil sorbent sorption capability tests

The most significant property of all sorbents is their ability to bind several times the amount of contaminant relative to their own weight, the sorption capability. The sorption capacity of the sorbent mainly depends on the size of the surface area to which the dangerous substance, oil product can adhere as well as surface type (surface structure). The size of the surface of the sorbents and the increase in their sorption capacity are influenced by various cracks such as pores and capillaries. The average surface area of the sorbent is expressed in units of area (m^2) per unit mass (g, kg). Sorbents have a surface area most often about 1,000,000 $m^2\ kg^{-1}$ [33, 34]. The sorption capacity expresses how many times the sorbent binds more liquid to its weight or how many grams of contaminant bind 1 g of sorbent.

The American Society of Testing and Materials unified the methods of testing the sorption capacity of sorbents—ASTM F 726: Standard Test Method for Sorbent Performance of Adsorbents [35]. This test method describes laboratory tests that describe the "Short-term Oil-Absorption Test" and other non-emulsion floating, non-emulsion liquids from the water level. The comparison of sorption capability different sorbent material is given in **Table 2**.

3.3.2 Water contamination after cleanup process

No cleaning process can be 100%. Various studies document that after mechanical cleaning, 10–15% of oil substances remain in the water [14, 32]. We made an experiment, where the samples of surface water (1 liter) were contaminated with diesel fuel and engine oil in amount of 1 g, 5 g, and 10 g. The layer of petroleum substance was built on the bottom of water. The oil film was stronger after application of engine oil. On the oil substances was applicated sorbent (expanded perlite) in the way to cover entire bottom of water level, which treated on oil substance. Contaminated sorbent was removed after 3 (diesel fuel) or 10 min (engine oil), and the amount of NES was determined in the water samples. The results are in **Table 3**.

Comparing the determined amounts of NES in "cleaned" water samples with legislative limits, we can say that all samples are still contaminated. The highest amount of NES was determined in the sample contaminated with diesel fuel (contamination with 10 $g.L^{-1}$). Amount of NES was in all samples higher than 0.1 $mg.L^{-1}$, what according to research of Hybská et al. could lead to toxicological treat of aquatic organisms in the first step and cumulation of oil substances in other environment components [25].

Sorbent/sorbent form	Oil kind	Oil sorption	Reference
Expanded perlite/granulate	motor oil	Up to 3 x	[26]
Expanded perlite/granulate	diesel fuel	Up to 4.5 x	[26]
Expanded perlite/granulate	gasoline	To 3 x	[34]
Expanded perlite/granulate	diesel fuel	Up to 3 x	[34]
Cellulosic waste/fiber material	gasoline	1.2 to 1.3 x	[34]
Cellulosic waste/fiber material	diesel fuel	1.7 to 1.9 x	[34]
calcium carbonate and clay/granulate	motor oil	1.4 x	[33]
Diatomaceous earth (moler) granulated, calcined	motor oil	1.1 x	[33]
Polypropylene/fiber	motor oil	14 x	[33]
Leaves residues	motor oil	7.4 x	[33]
Peat moss	motor oil	Up to 6 x	[33]
Needles	motor oil	4.4 x	[33]
Saw dust	motor oil	Up to 3.3 x	[33]
Urethane-isocyanate-alcohol polymer/granulate	motor oil	34.4 x	[32]
Polypropylene/fiber web	crude oil	10 x	[32]
Exfoliated graphite	crude oil	80 x	[32]
Cellulose	crude oil	Up 20 x	[32]
CF3-functionalized silica aerogel	crude oil	Up to 237 x	[32]
Acetylated rice straw	motor oil	Up 20 x	[32]

Table 2.
Comparison of different sorbent materials.

Contaminant	Sorbent	Added amount of contaminant ($g.L^{-1}$)	Amount of NES after contamination ($mg.L^{-1}$)	Amount of NES after clean up ($mg.L^{-1}$)
Engine oil	Expanded perlite	1	795	0.474
		5	3.980	3.569
		10	9.152	19.77
Diesel fuel		1	680	17.55
		5	4.370	24.13
		10	8.525	25.84

Table 3.
Determined NES amount in the water samples before and after sorbent treatment [26].

3.3.3 Absorbent sock (absorbent snake)

The sorption sock, called also sorption snake (**Figure 7**), is a combination of the containment function (the oil) booms and function of the oil contamination collecting function (sorbents).

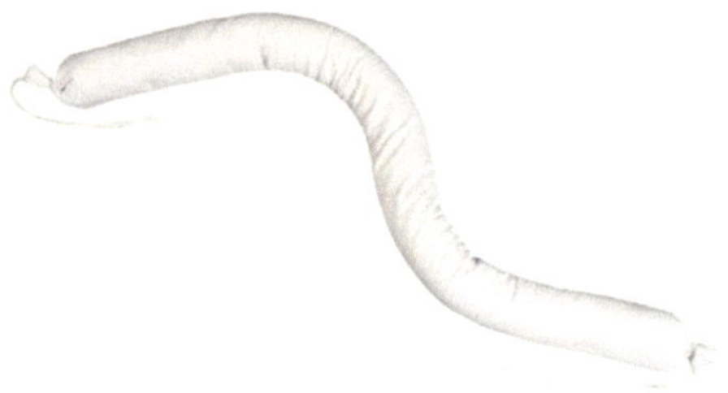

Figure 7.
The absorption sock (absorption snake) [36].

Figure 8.
The use of the absorption socks on solid surface (left) and on the water surface (right) [37].

The sorption snakes are intended for solving minor of oil spills—rivers and streams, lakes, ponds, etc., and also for use on solid surfaces (**Figure 8**). These are means intended for manual operations. Their primary function is to contain the oil spill so that the contamination does not spread to a larger area or to the shores similar to oil containment booms.

We can meet sorption socks in various designs—longer with a small diameter or on the contrary, a shorter larger diameter. The sorption snakes consist of two types of adsorbents. Adsorbent I forms the textile cover of the sorption sock, and adsorbent II represents the filling. Any loose sorbent can be used as a filler.

In our experiment, we tested the sorption capacity of the sorption sock according to the ASTM F 726 methodology. The cover was tested separately as adsorbent I and the filling as adsorbent II. Subsequently, the entire absorption snake was also tested. All materials were tested for the absorption of oil and water/oil mixtures. The results are shown in **Table 4**.

From the results, it seems at first sight that the sorption capacity of the absorbent sock as a whole is low on average. However, it is necessary to realize the primary role that the sorption sock fulfills—the containment of the oil spill. Oil recovery is an associated task.

It is also necessary to remember that neither the cover nor the filler is in contact with the contaminant in its entire volume. As can be seen in **Figure 9**, about half of the

Sorbent	Sorption capacity g of oil/g of sorbent crude oil	Sorption capacity g of oil/g of sorbent oil + water mixture
Textile (cover)	7.15 ± 0.47	6.50 ± 0.17
Granulate (filler)	4.6 ± 0.42	5.83 ± 0.30
Absorption sock	1.67 ± 0.13	1.06 ± 0.09

Table 4.
Results of sorption capacity test of absorption sock.

Figure 9.
The absorption sock after absorption of crude oil (left) and oil from the water/oil mixture(right).

sorption sock was soaked with oil. This can be considered as excellent results, which show that the use of an absorption sock for small oil spill disposal is a very suitable choice.

The combination of mechanical methods (oil containment booms, oil skimmer, or sorbents) of cleaning oil spills is one of the most used and also the most effective [2, 14]. Complications for the effective operation of booms, skimmers, and sorbents are adverse weather (strong wind and rain), strong waves and currents. These significantly affect oil spill behavior and also make difficult operations with the equipment. Commonly used booms operate at a maximum wave speed of 7 knots. At higher speeds, it is necessary to deploy special types of booms as well as special equipment [14]. Otherwise splash-over or sinking of the boom may occur. It is very important to choose the right type of containment boom for using, because the boom is the first deployed device on the site and the last one, which will be remoted. Depending on the current wind-, waves-, and currents conditions and the size of the oil spill, the type of sorbent to be used is also chosen. Particles that are too light can be blown into the open sea by strong winds, and on the contrary, particles that are too heavy could settle to the bottom together with bounded oil particles. Adverse conditions such as wind and waves, in addition to complicating the operations with the equipment, significantly increase the economic costs of oil spill cleaning. Despite these disadvantages, mechanical methods of oil spill cleanup are among the most widely used. Advantages include the possibility of recovery of captured oil, not introducing additional chemicals into the environment, the possibility of combination with biotechnological methods [2, 3, 14, 27, 38].

Alternatives to mechanical methods are in-situ burning, chemical dispersion, biological methods. In-situ burning is suitable for rapid burning of oil on the water surface in case of large spills. The disadvantage is the production of a huge amount of smoke containing toxic substances. These can spread over large areas in windy weather and affect light conditions and air quality in them. During chemical dispersion, surfactant substances are sprayed onto the oil layer, disrupting it, reducing the surface tension of water, which leads to emulsification and dispersion of oily substances throughout the water column. A major disadvantage of both mentioned methods is the secondary pollution in the vicinity of the accident, which contributes to the damage of natural aquatic and terrestrial ecosystems [14, 27, 38].

Against the background of these facts, mechanical methods of removing oil spills are considered highly effective and also advantageous from an environmental point of view [2, 14, 38]. Since it is not entirely possible to prevent oil spill accidents, the research and development of these methods is one of the challenges of many scientists. Improvement occurs mainly in the development of new or modified sorbents in order to increase their sorption capacity, the ability to recover oil, ease of handling. Cost reduction and environmental suitability are also important factors. Organic and synthetic materials are investigated (multi-layered graphene, chitosan, kenaf fiber, ZnO), waste material is used, surface treatment of the sorbent, creation of composite sorbents with the addition of nanoparticles, additive substances inducing the magnetic properties of the sorbent, etc. [2, 3, 14, 27, 39–43].

4. Conclusions

Oil is an important industrial commodity worldwide. Although there is a lot of talk about replacing it in the context of climate change, its use is unlikely to stop in the next few years. Its extraction, transport, and use are associated with leaks into the environment during accidents or during normal operation. Petroleum substances have a very harmful effect on the environment, affect the oxygen regime in water and soil, are toxic to aquatic organisms and animals as well as to humans.

The chapter deals with mechanical methods for oil spill removal. The first part briefly describes the interactions of oily substances in the water environment, the distribution of oil particles in water bodies, their harmful effects and ecotoxicity. The main part deals with the possibilities of cleaning oil spills using physical (mechanical) methods. The use of oil containment booms, oil skimmers, and sorbents in accidental oil spills solutions is explained and described.

Based on research, experiments, and studies, the effectiveness of several types of sorbents is compared. The determination of nonpolar hydrocarbons in water after cleanup process proved that these cleaning procedures must be combined with other methods, because the water does not meet the limit requirements for surface water. Although the removal of oil pollution was very effective, not a single water sample met the criteria for discharge into the recipient. The chapter also describes testing the sorption capacity of absorption sock, which can be used for smaller oil spills removal, because it is advantageous to combine two functions—containment of oil and recovery of oil compounds.

The advantages and disadvantages of individual cleanup methods are summarized too. Although development and progress in the use of new procedures, innovative technical equipment and devices, progressive materials, and cleanup process methods are still progressing, the best way for humanity and the environment is to prevent and avoid oil spills and at the same time be as well prepared as possible to deal with accidents.

Acknowledgements

This work was supported by the Slovak Research and Development Agency under the contract No. APVV-17-0005 (50%) and Scientific Grant Agency VEGA, contract No. 1/0454/20 (50%).

Author details

Veronika Veľková[1*], Helena Hybská[2] and Tatiana Bubeníková[3]

1 Department of Fire Protection, Technical University in Zvolen, Zvolen, Slovakia

2 Department of Environmental Technology, Technical University in Zvolen, Zvolen, Slovakia

3 Department of Chemistry and Chemical Technology, Technical University in Zvolen, Zvolen, Slovakia

*Address all correspondence to: veronika.velkova@tuzvo.sk

References

[1] Doshi B, Repo E, Heiskanen JP, Sirvio JA, Silanpää M. Sodium salt of oleolyl carboxymethyl chitosan: A sustainable adsorbent in the oil spill treatment. Journal of Cleaner Production. 2018;**170**:339-350. DOI: 10.1016/j.jclepro.2017.09.163

[2] Tan JY, Low SY, Ban ZH, Siwayanan P. Oil spill clean-up with bio-sorbents. BioResources. 2021;**16**:8394-9416. DOI: 10.15376/biores.16.4.Tan

[3] dos Anjos RB, Hilário LS, dos Anjos ASD, de Araújo Costa ECT, Frota TMP, Gondim AD, et al. Combined treatment (alkali + thermal) of Calotropis procera fiber for removal of petroleum hydrocarbons in cases of oil spill. Polymers. 2021;**13**:3285. DOI: 10.3390/polym13193285

[4] Mohit K: 11 Major Oil Spills of The Maritime World. [Internet]. 2021. Available from: https://www.marineinsight.com/environment/11-major-oil-spills-of-the-maritime-world/

[5] Moss L: The 14 biggest oil spills in history. [Internet]. 2022. Available from: https://www.treehugger.com/the-largest-oil-spills-in-history-4863988

[6] Marghany M. Automatic mexico gulf oil spill detection from radarsat-2 SAR satellitebdata using genetic algorithm. Acta Geophysica. 2016;**64**:1916-1941. DOI: 10.1515/acgeo-2016-0047

[7] Barbier CJ. MDL – 2179 Oil Spill by the Oil Rig "Deepwater Horizon". [Internet]. 2017. Available from: https://www.laed.uscourts.gov/OilSpill/OilSpill.htm

[8] Enujiugha VN, Nwanna LC. Aquatic oil pollution impact indicators. Journal of Applied Sciences & Environmental Management. 2004;**8**:71-75

[9] Deepwater horizon oil spill. [Internet]. 2017. Available from: https://oceanservice.noaa.gov/news/apr17/dwh-protected-species.html

[10] Effects of oil pollution on marine wildlife. [Internet]. Available from: http://oils.gpa.unep.org/facts/wildlife.htm

[11] Al Bayaty HJA, Alsultan ATZ. Detection of oil spill pollution on water surface using microwave remote sensing techniques. Materials Science and Engineering. 2020;**737**:012249. DOI: 10.1088/1757-899X/737/1/012249

[12] Yan Z, Fu H, Wang Z, Sun B, Ren J. The effect of plunging water of waves on the entrainment of spilled oil at sea. Water Resources. 2021;**48**:41-47. DOI: 10.1134/S0097807821010280

[13] Pan Y, Jia Y, Wang Y, Xia X, Guo L. Study on diesel vertical migration characteristics and mechanism in water-bearing sand stratum using an automated resistivity monitoring system. Environmental Science and Pollution Research. 2018;**25**:3802-3812. DOI: 10.1007/s11356-017-0698-3

[14] Mullin J. Technology Assessment & Research (TA&R) project categories: Mechanical containment and recovery. [Internet]. 2010. Available from: https://web.archive.org/web/20100507135220/http://www.mms.gov/tarprojectcategories/mechanic.htm

[15] Passow U, Ziervogel K, Asper V, Diercks A. Marine snow formation in the aftermath of the deepwater horizon oil spill in the Gulf of Mexico.

Environmental Research Letters. 2012;**7**:035301. DOI: 10.1088/1748-9326/7/3/035301

[16] Hybská H, Samešová D, Ollerová H: Impact of Organic Pollutants on the Environment. 1st ed. Zvolen: Technical University in Zvolen. 2015. p. 187. ISBN 978-80-228-2766-9

[17] Shogren RL, Petrovic Z, Liu Z, Erhan SZ. Biodegradation behavior of some vegetable oil-based polymers. Journal of Polymers and the Environment. 2004;**12**:173-178. DOI: 10.1023/B:JOOE.0000038549.73769.7d

[18] Saval S. Bioremediation: Clean-up biotechnologies forsoils and aquifers. In: Environmental Biotechnology and Cleaner Bioprocesses. London: Taylor and Francis Limited; 2000. pp. 155-166

[19] Mariano AP, Kataoka APAG, Angelis DF, Bonotto DM. Laboratory study on the bioremediation of diesel oil contaminated soil from a petrol station. Brazilian Journal of Microbiology. 2007;**38**:346-353. DOI: 10.1590/S1517-83822007000200030

[20] Lopes PRM, Montagnolli RN, Domingue RF, Bidoia ED. Toxicity and biodegradation in sandy soil contaminated by lubricant oils. Bulletin of Environmental Contamination and Toxicology. 2010;**84**:454-458. DOI: 10.1007/s00128-010-9945-8

[21] Tamada IS, Lopes PRM, MontagnolliRN, BidoiaED. Biodegradation and toxicological evaluation of lubricant oils. Brazilian Archives of Biology and Technology. 2012;**55**:951-956. DOI: 10.1590/S1516-89132012000600020

[22] Cecutti CH, Agius D. Ecotoxicity and biodegradability in soil and aqueous media of lubricants used in forestry applications. Bioresource Technology. 2008;**99**:8492-8496. DOI: 10.1016/j.biortech.2008.03.050

[23] Borodulina TS, Polonskiy VI, Vlasova ES, Shashkova TL, Grigor YS. Effect of oil-pollution of water on delayed fluorescence of the algae *Chlorella vulgaris Beijer* and survival rate of the cladoceran *Daphnia magna Str*. Contemporary Problems of Ecology. 2011;**4**:107-111. DOI: 10.1134/S1995425511010139

[24] Martinez JF, Villaseńor R, Rioe G, Chaves EF. Toxicity of the crude oil water soluble fraction and kaolin – adsorbed crude oil on *Daphnia magna (Crustacea: Anomopode)*. Archives of Environmental Contamination and Toxicology. 2008;**48**:444-449

[25] Hybská H, Mitterpach J, Samešová D, Schwarz M, Fialová J, Veverková D. Assessment of ecotoxicological properties of oils in water. Archives of Environmental Protection. 2018;**44**:31-34. DOI: 10.24425/122300

[26] Veľková V, Hybská H, Bubeníková T. The effectivity of selected sorbents used to the oil leaks disposal. Waste forum. 2018;**3**:307-313

[27] Balauka YA, Mayorava KI, Ayoub Z. Emergency sorbents for oil and petroleum product spills based on vegetable raw materials. Materials Science and Engineering. 2018;**11**:451-457. DOI: 10.1088/1757-899X/451/1/012218

[28] Available from: https://www.udrzba.cz/sorbenty-a-havarijni-prostredky/olejove-sorbenty-havarijni-soupravy-uklidove-sorbenty/norne-steny-spc/env-150-v-sorpcni-koberec-hydrofobni-60.html

[29] US EPA, OLEM skimmers. [Internet]. Available from: https://www.epa.gov/emergency-response/skimmers

[30] Weir oil skimmer. [Internet]. Available from: https://www.desmi.com/products-solutions-library/tarantula-oil-skimmer

[31] Drum oil skimmer. [Internet]. Available from: http://www.aaradhyaenvirotech.in/wastewater-treatment-plant-3995349.html

[32] Adabajo MO, Frost RI, Kloprogge JT, Carmody O. Porous materials for oil spill cleanup: A review of synthesis and absorbing properties. Journal of Porous Materials;**10**:159-170. DOI: 10.1023/A:1027484117065

[33] Mojzis M, Bubenikova T, Zachar M, Kacikova D, Stefkova J. Comparison of natural and synthetic sorbents' efficiency at oil spill removal. BioResources. 2019;**14**:8738-8752. DOI: 10.15376/biores.14.4.8738-8752

[34] Bubeníková T, Veľková V, Čerevka M: The effect of some sorptive materials at the disposal of oil substances leakage from water level. In: Advances in Fire and Safety Engineering. Trnava: Alumni Press; 19-20 oct 2017. pp. 111-116. ISBN 978-80-8096-245-6

[35] ASTM F 726: Standard test method for sorbent performance of adsorbents

[36] Absorption sock. [Internet]. 2018. Available from: https://www.reoamos.sk/hydrofobny-sorpcny-had-plneny-sorpcnou-drvinou/d-6146/

[37] Absorption snake. [Internet]. 2018. Available from: https://www.reoamos.sk/hydrofobny-sorpcny-hladinovy-koberec-115-x-04-m/d-10316/

[38] Torres DHA, da FC D, Bahiana DR, Haddad AN, Chinneli Ch K, Soares CAP. Oil spill simulation and analysis of its behavior under the effect of weathering and chemical dispersant: A case study of the Bacia de Campos—Brazil. Water, Air, and Soil Pollution. 2020;**231**:521. DOI: 10.1007/s11270-020-04857-8

[39] Onwuka JCH, Agbaji EB, Ajibola VO, Okibe FG. Treatment of crude oil-contaminated water with chemically modified natural fiber. Applied Water Science. 2018;**8**:86. DOI: 10.1007/s13201-018-0727-5

[40] Akpomie KG, Conradie J. Enhanced surface properties, hydrophobicity, and sorption behavior of ZnO nanoparticle–impregnated biomass support for oil spill treatment. Environmental Science and Pollution Research. 2021;**28**:25283-25299. DOI: 10.1007/s11356-021-12451-6

[41] Vocciante M, Finocchi A, de D'Auris AF, Conte A, Tonziello J, Pola A, et al. Enhanced oil spill remediation by adsorption with interlinked multilayered graphene. Materials. 2019;**12**:2231. DOI: 10.3390/ma12142231

[42] Mehmood A, Khan FSA, Mubarak NM, Mazari SA, Jatoi AS, Khalid M, et al. Carbon and polymer-based magnetic nanocomposites for oil-spill remediation—A comprehensive review. Environmental Science and Pollution Research. 2021;**28**:54477-54496. DOI: 10.1007/s11356-021-16045-0

[43] Ezzat AO, Ali MS, Al-Lohedan HA. Synthesis, characterization, and application of magnetite nanoparticles coated with hydrophobic polyethyleneimine for oil spill cleaning. Hindawi Journal of Chemistry. 2022;**2022**:1-10. DOI: 10.1155/2022/3368298

Chapter 4

On the Possibility of Non-Local and Local Oil Spills Striking the Shores of North Carolina and South Carolina

Leonard J. Pietrafesa, Shaowu Bao, Paul T. Gayes and Farid Askari

Abstract

Oil spills, the releases of liquid petroleum hydrocarbons into the marine environment, have occurred in the Gulf of Mexico (GOM) of the United States (U.S.A). However, no oil spills have ever affected the Eastern Atlantic Seaboard (EAS) of the U.S.A. Nonetheless, we demonstrate from data and numerical modeling that oil spills in the GOM have the potential to reach the U.S.A. EAS via a combination of atmospheric storms, major ocean currents and atmospheric wind driven surface currents. The basis for this hypothesis is that in August of 1987, a Karena Brevis toxin plant outbreak occurred in the GOM, and several weeks hence, showed up on the shores of North Carolina and South Carolina. We recreate that environmental scenario employing atmospheric and oceanic data from 1987, Sea Surface Temperature (SST) images, and via numerical modeling, that an atmospheric cold front, the combination of the Loop Current, the Florida Current, and Gulf Stream Frontal Filaments, created the pathways for the transport of K-Brevis plants from the Gulf to the U.S.A. EAS. Numerical model output of oil spill scenarios, both non-local in the GOM and local to the Carolinas, is presented to prove that this latter hypothesis has credibility and viability.

Keywords: cold fronts, K-breve, red tide, Loop Current, Gulf Stream, frontal filaments, mid-latitude cyclones

1. Introduction

Oil Spills have never been reported as having occurred along the coasts of either South Carolina (SC) or North Carolina (NC). However, we present evidence, by way of a surrogate to oil droplets; a marine-based toxic biological plant, where a non-local source invasion of oil could occur in the coastal waters of those states. Additionally, via numerical modeling, we show that both non-local and local spills could invade SC and NC beaches. By way of example, an industrial disaster, the British Petroleum

Oil (BPO) Company Deepwater Horizon oil spill, occurred in late April 2010 in the eastern Gulf of Mexico (GOM). The BPO vessel, the Macondo Prospect, sprung a leak that resulted in the largest marine oil spill in the history of the petroleum industry. In truth, the BPO oil spill was contained in the GOM. Therefore, a presumption could be that given the BPO spill in the GOM, that future spills will be contained in kind. That is not the case, given evidence provided by our surrogate GOM-based toxic marine plant.

On October 31, 1987, Onslow Bay, NC continental shelf waters became infested with a yellow-green toxic organism. Investigators from the National Oceanic & Atmospheric Administration (NOAA) National Marine Fisheries Service (NMFS) Laboratory in Beaufort NC (Dr. P. Tester, p.c.) determined that the yellow-green patches consisted of the one-celled plant organism Ptychodiscus Brevis or the Karina Brevis Dinoflagellate, and accompanying marine algae indigenous to tropical waters such as the GOM. The effects of the one-celled intruder were immediate and widespread. Shellfish, such as clams, scallops and oysters became infected and rendered inedible. This particular Red Tide organism contained a neurotoxin that affected the nervous systems of higher life forms, including humans. As the neurotoxins become airborne via breaking waves, beachcombers and surf fishers suddenly felt the sensations of burning eyes and lungs, nausea, and dizziness. Subsequently shell fishing was banned in NC and the beaches were closed.

Federal and university scientists became suspicious that the Red Tide dinoflagellate was transported to the subtropical waters of NC and SC from the tropical south. If so, then what was the source of the K-Brevis and what was its pathway? The answer was addressed by Pietrafesa et al. [1] in which it was hypothesized that an atmospheric cold front in the GOM in August, the Loop Current, the Gulf Stream, and the atmospheric wind field of NC in late September and early October created a hypothetical scenario for the realization of this event. Could it happen again? Possibly. Further, could GOM spilled oil, a non-local event for the Carolinas, be transported the same way as the Red Tide to NC and SC? Further, could oil spilled locally off the NC coast be transported to NC and SC beaches and estuaries? Conventional wisdom is that the oil would be swept to away by the Gulf Stream and distributed across the North Atlantic Ocean to the north. However, via numerical model experiments we show that NC spills could reach the beaches of NC and SC. We develop the physical descriptive and numerical modeling scenarios below.

In Section 2, we describe the 1987 eastern GOM Red Tide event that reached NC and SC beaches. In Section 3, we discount the Astronomical Tides as a potential cause of non-local or oil spills reaching the NC or SC beaches. In Section 4, we describe Gulf Stream Variability and Frontal Filaments. In Section 5, we revisit the 1987 Red Tide event that invaded NC and SC coastal waters, via satellite imagery and then numerically model the non-local Red Tide event with modern numerical modeling. In Section 6, we model hypothetical local oil spills on the NC coast during the passages of a GSF and a typical mid-latitude cyclone. Section 7 includes the conclusions and summary.

2. The 1987 Eastern Gulf of Mexico Red Tide Event

On August 24, 1987, a breakout of the Red Tide was reported off the coast of Naples, Florida, in the GOM. As the oceanic currents flow, Naples is approximately 1600 kilometers (1010 miles), from the coasts of the Carolinas. If the Red Tide plants were able to jump aboard the Loop Current in the Gulf of Mexico, they could have been transported down the west Florida coast, around the Florida Keys, through the Florida Straits where it would have become part of the westward flowing Antilles Current and then loaded into the northward flowing Florida Current, which then becomes the northward flowing Gulf Stream (**Figure 1**). From the east coast of Florida, the organism would have had to have traveled north reaching Onslow Bay NC and Long Bay NC/SC outer shelf waters sometime in early October. Let us first consider the conditions that were present at the time.

In **Figure 2a**, the NOAA National Weather Service (NWS) atmospheric pressure map for August 24, 1987 shows a high-pressure center located in the southeastern USA. The winds associated with this weather system on the west Florida shelf would have been to the south, thereby effecting an offshore transport of surface waters via an Ekman surface layer [3] from the shelf into the eastern side of the GOM Loop Current. Two weeks hence, we find a low-pressure center or atmospheric cyclone located in the southeast, as shown in the September 7 weather map in **Figure 2b**. The winds are to the north on the eastern Atlantic Florida shelf, thereby driving surface coastal waters offshore, again via Ekman surface layer dynamics [3, 4], into the Gulf Stream. Therefore, while the Red Tide organisms were likely in the area of the east Florida shelf, winds were unfavorable for onshore transport out of the western edge of the Gulf Stream and onto the shelf. Therefore, the organisms stayed in the Gulf Stream, on its western side, marching northward. They could have gone from the western side of the Gulf Stream via the Astronomical Tides or due to Gulf Stream–related phenomena. We investigate that further below in Sections 3, 4, and 5.

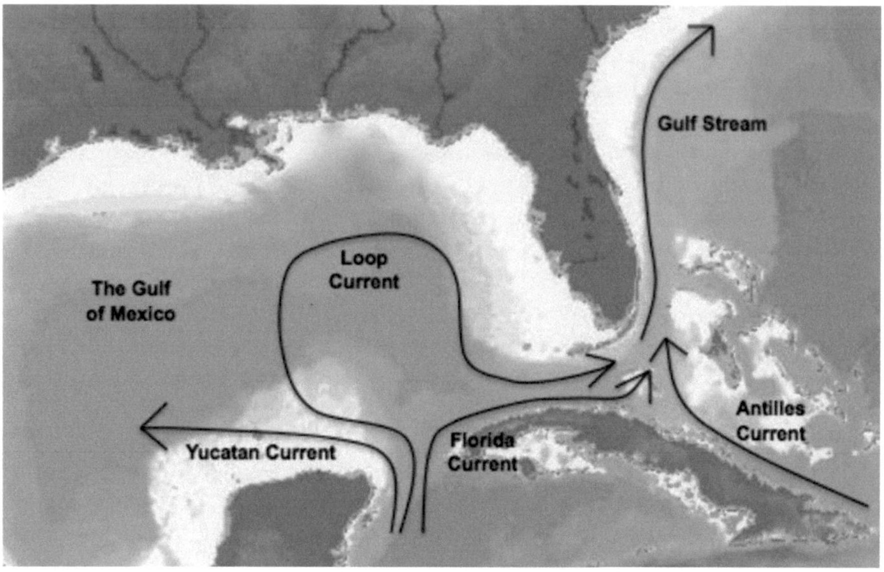

Figure 1.
The Loop, Antilles and Florida Currents, and the Gulf Stream (from [2]).

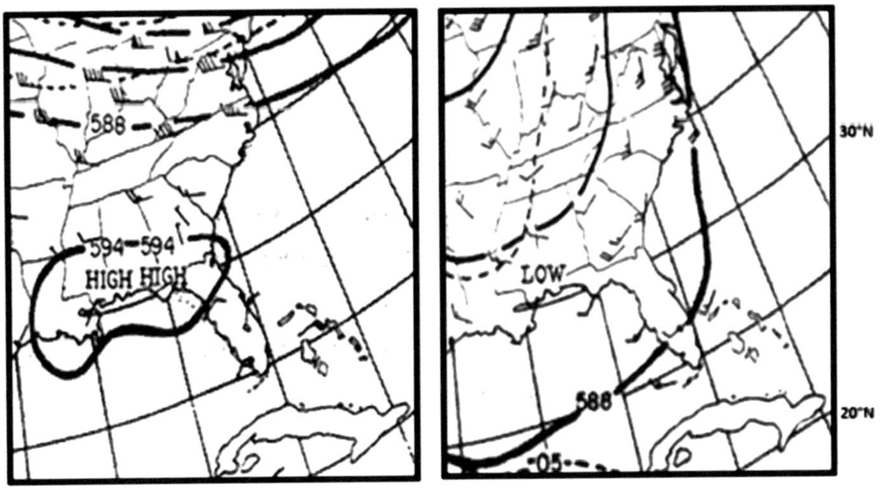

(a) (b)

Figure 2.
NWS 500 mb surface pressure maps. (a) Left image is August 24, 1987; (b) right image is September 8 1987. These maps are hard copies of those produced by the NWS and are copied with full fidelity.

3. Astronomical tides in the Southeast Atlantic

The astronomical tides of the Atlantic Ocean have been explored since ancient times. As early as 600 CE, medieval monks documented tidal changes throughout the coast of England, and they properly grasped the link between tides, the location of the sun, and the phases of the moon. The utilization of precise tidal gauges for continuous gathering data, as well as advanced computers for modeling and prediction, has greatly increased tide table accuracy and knowledge of the numerous constituent forces that shape and influence tidal behavior.

The tide in the S-shaped north-south Atlantic basin may indeed be conceived of as a unique phenomenon that acts like a massive standing wave traveling across the basin. A variety of complicated elements govern the pace, path, size, and behavior of the Atlantic tide, involving shoreline unusual features, seafloor topography, and dynamical patterns of wind and current. The most frequent and prominent tidal variety in the Atlantic Ocean Basin is the semidiurnal, which has two high and two low tides every tidal day (lasting about 24 hours and 50 minutes). Semidiurnal tides occur over the entire eastern edge of the Atlantic, as well as across the majority of North and South America. Mixed tides, or those with both diurnal (one high and one low tide per day) and semidiurnal oscillations, predominate in the Gulf of Mexico and the Caribbean Sea, as well as along the southeastern coast of Brazil and Tierra del Fuego, in some areas of the Mediterranean, and along the coast of Labrador; the only purely diurnal tides occur in portions of the Gulf of Mexico.

Tidal dynamics for the Southeast Atlantic continental margin have been thoroughly discussed in Pietrafesa et al. [5]. As such, the tide on the NC and SC coasts consists of two principal constituents, the near semi-diurnal, M2, with a principal period of 12 hours, and 25 minutes and the diurnal, S1, of 24 hours. According to that study, the M2 and S1 tides are both Poincare Waves. The net result is that a parcel of water subjected to only the tides would traverse clockwise around an ellipse with a major onshore-offshore axis of 2 km and a minor alongshore axis of 1 km. The net

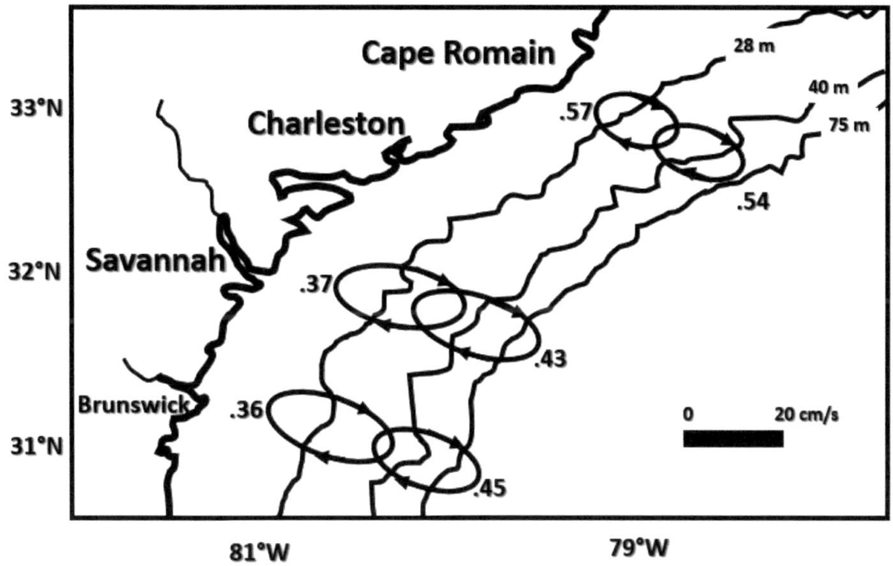

Figure 3.
Water particle motions due to the M2 astronomical tides along the 28 m and 40 m isobaths off Georgia, South Carolina, and North Carolina. The calculations of the clockwise rotating water parcels are computed directly from Eulerian current meter observations [5]. The numbers on the ellipses indicate the ratios of the onshore/offshore ellipse axes to the alongshore axes. The ellipse axes are about 2 km onshore/offshore and 1 km alongshore.

result would be that every 12 hours and 25 minutes, a parcel of water would end back up where it started. These observed water particle motions are visualized in **Figure 3**. Therefore, the astronomical tide is discounted as having been the agent responsible for moving the dinoflagellates across the shelf. We next consider Gulf Stream variability and features.

4. Gulf stream variability and frontal filaments

In a Sea Surface Temperature (SST) NOAA GOES satellite-based study, Pietrafesa [6] reported that the Gulf Stream Current deflects offshore near 31°N, 79°W, meanders laterally thereafter, and its lateral meander variability decreases downstream of this deflection. The seaward deflection of the Gulf Stream Current was determined to be caused by the presence of a topographic irregularity, which became known as the "Charleston Bump," actually Hoyt's Hill in the geological history of the region. The study conjectured that the topographic feature changed the vertical vorticity of the Gulf Stream by shrinking the vertical water column such that the Gulf Stream had to move offshore to deeper water to preserve its angular momentum balance. After the Gulf Stream moves into deeper water, given angular momenta requirements, it reroutes itself toward the coast. The process was speculated to affect the generation of Topographic Rossby Waves (TRWs) in that locale by Rooney et al. [7] and Pietrafesa and Janowitz [3]. These waves were found to propagate to the north along the shelf breaks of SC and NC [3, 8, 9], with periods between 2 and 12 days and propagation speeds of 30–40 km/d. The study of Sun and Pietrafesa [9] also discovered that the Gulf Stream Front has an inherent 8-day baroclinic instability that is a persistent source of downstream propagating waves. John and Schott [10] staged an Eulerian

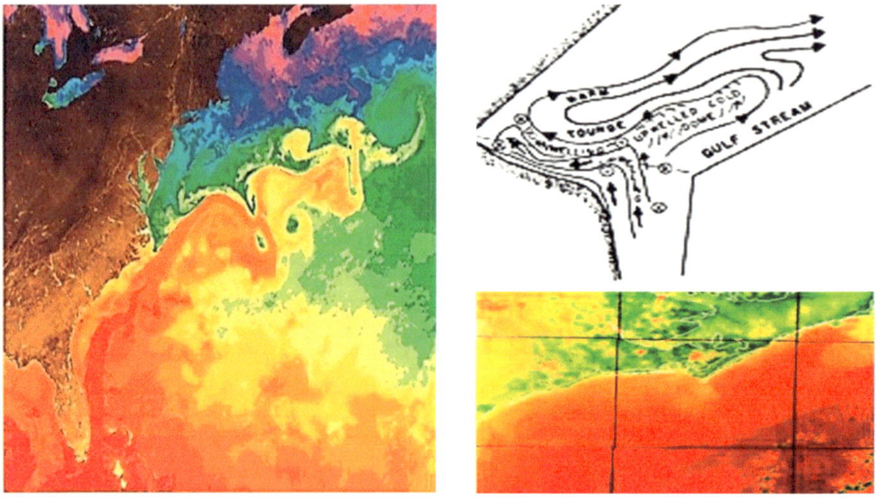

(a) (b)

Figure 4.
(a) Left panel depicts the Gulf Stream and its configurations in the North Atlantic Ocean basin; (b) right panel shows a Gulf Stream frontal filament in a NOAA SST mage and its conceptual flow field.

current meter study on the FL outer shelf and determined that the Gulf Stream meandered laterally, and they concluded that these onshore-offshore motions were northward propagating waves, with dominant wavelengths of 340 km and 170 km, periods of 12 days and 5 days, and propagation speeds of 28 km/d and 36 km/d, respectively. A NOAA GOES SST image is shown in **Figure 4a**, which shows Gulf Stream variability along the Atlantic Seaboard. We note the offshore deflection off Charleston and a variety of frontal features. The horizontal crests of these laterally meandering waves can bring surface layer parcels of water onto the outer continental shelves of the coasts of NC and SC. Moreover, these waves can fold back at their crests.

Pietrafesa and Janowitz [11] and Pietrafesa [12] evaluated Eulerian current meter data off NC and FL, respectively, and provided a detailed current meter–based spatial and temporal map of a TRW meander crest that folded back, which they referred to as a Frontal Filament, wrapped around an offshore cold core eddy (**Figure 4b**). In summary, meanders and the frontal filaments and the eddies they generate serve as the principal form of mesoscale variability along the path of the Gulf Stream Current, on the outer continental shelf within the South Atlantic Bight (between Cape Canaveral, FL, and Cape Hatteras, NC). From this suite of comprehensive studies, the Gulf Stream has been shown to display many degrees of freedom (**Figure 4a**). **Figure 4b** is a beautiful SST representation of a GSF. Warm Gulf Stream water traveling in the crest of a Meander folds back onto the outer shelf and then travels southwestward into and around the filament and then turns toward the northeast and back into the Gulf Stream Front. The Cartoon in **Figure 4b** shows that pictorially.

5. The 1987 Red Tide event as viewed in SSTs and numerical model output

In **Figure 5**, employing NOAA AVHRR-derived Sea Surface Temperature images (SST), courtesy of Dr. Steven Baig of NOAA's Atlantic & Oceanographic Laboratory

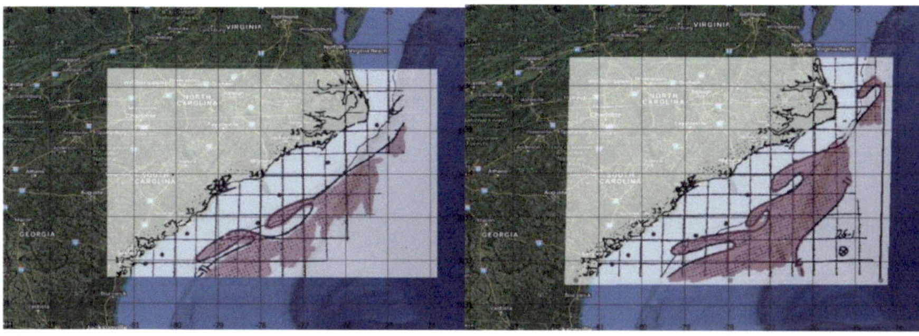

Figure 5.
NOAA AVHRR imagery of Gulf Stream waters with frontal filaments. (a) Left panel, October 05, 1987, image showing two filaments; (b) October 09, 1987, image showing three filaments. (SST images were provided by Dr. S. Baig, AOML). The magenta coloring is employed to depict the location of the Gulf Stream and its Frontal Filaments.

(AOML) that outlined the Gulf Stream and its frontal features, we see the presence of GSFs. At the time, these SST maps were a product that was hand drawn by Dr. Baig of AOML as a public service particularly the fishing industry. In the upper panel, two GSFs are shown to have been located between 32° and 33.5° on October 05. In the **Figure 5** lower panel, a third GSF has appeared and the three frontal features are between 32.25° and 34.25°. The southernmost GSF in **Figure 5** upper panel moved to Onslow Bay offshore waters as shown in **Figure 5** lower panel. An additional GSF had by then formed east of Cape Romain, SC. For the GSF located offshore of Onslow Bay to have propagated there from its previous location offshore of Charleston, SC, it would have had a phase speed of approximately 42 cm/sec or 36 km/day. If this phase speed of propagation is representative of the speed of parcel movement along the western wall of the Gulf Stream, the Gulf Stream frontal zone, then it would have taken a patch of water and its constituents 45 days to go from Naples, Fla., to Onslow Bay. We will test this with our numerical model scenario present below.

From the above data-based hypothesized description, Red Tide dinoflagellates could have been loaded into the Loop Current in the Gulf of Mexico offshore of Naples, Fla., and eventually could have been positioned in a large Gulf Stream Frontal Filament (GSF) offshore of Onslow Bay on Oct. 09. If the dinoflagellates were located in surface layer waters of the GSF, then the obvious question occurs: How could the dinoflagellates have been transported out of the filament and across Onslow Bay, a distance of 90–110 km (56–68 mi.) by October 31 when the Red Tide was first observed on the NC beaches? To address this question, we must ask an additional one. What physical processes could exist at this time of year that would move one-celled, microscopic drifters across the width of Onslow and Long Bays? We next employ a numerical atmospheric and ocean current model system to simulate the events of 1987.

In our numerical model simulation, we employ two different reconstructed wind fields, so as not to appear to bias the atmospheric driving force. On the oceanic side, we employ the Regional Ocean Model System (ROMS) ocean circulation model [13]. **Figure 6** depicts the seeding of coastal waters on the West FL shelf with Red Tide plants that are assumed passive tracers. Two model simulations were conducted to offer comparisons. One simulation employed the North American Regional Reconstruction (NARR) winds (https://www.emc.ncep.noaa.gov›mmb›rreanl›index.

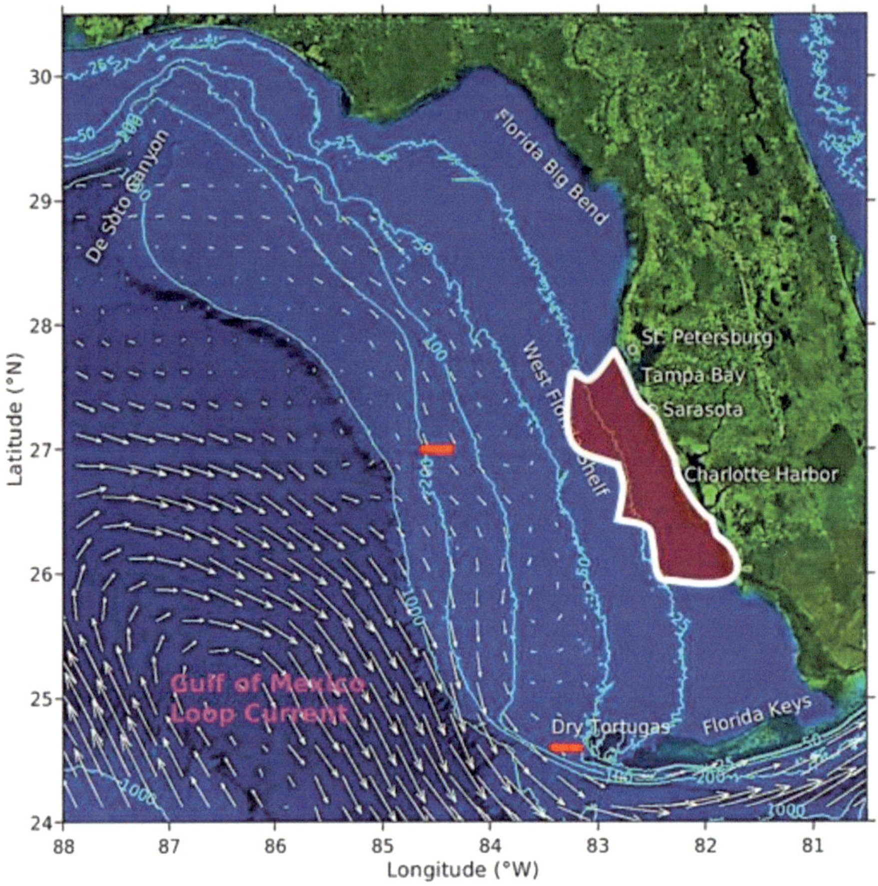

Figure 6.
Patch of red tide cells dumped into the ROMS Ocean current model in the surface waters of the west coast of FL on August 24, 1987, with a triple-nesting approach in the ROMS Ocean currents model.

html) and the second, the European Centre for Medium-Range Weather Forecasts (ECMWF) winds (https://www.ecmwf.int).

Figure 7 presents the NARR wind fields used in the model experiment, by way of example. As the winds on the west FL shelf were consistent from late August to early September, we present that sequence. In **Figure 8**, the numerical model simulations employing the NARR winds versus the ECMWF winds (which are not shown) are presented. They are quite comparable, providing credence to our mechanical wind-forcing hypothesis.

In the numerical model experiments, two conclusions are reached: (1) the atmospheric winds in late August and early September 1987 on the West Florida Shelf were sufficient to transport the Karina Breve cells from the west coast of Florida to the Loop Current to the Florida Current—and into Gulf Stream system; (2) the combination of the wind effects, from Cold Fronts in the GOM and ETCs; on the other hand, the effect of the Gulf Stream meanders and frontal filaments are necessary to transport the passive tracer from the Gulf Stream to the inshore area of the NC and SC coasts. We next consider the atmospheric wind fields along the southeastern Atlantic Seaboard in the fall 1987.

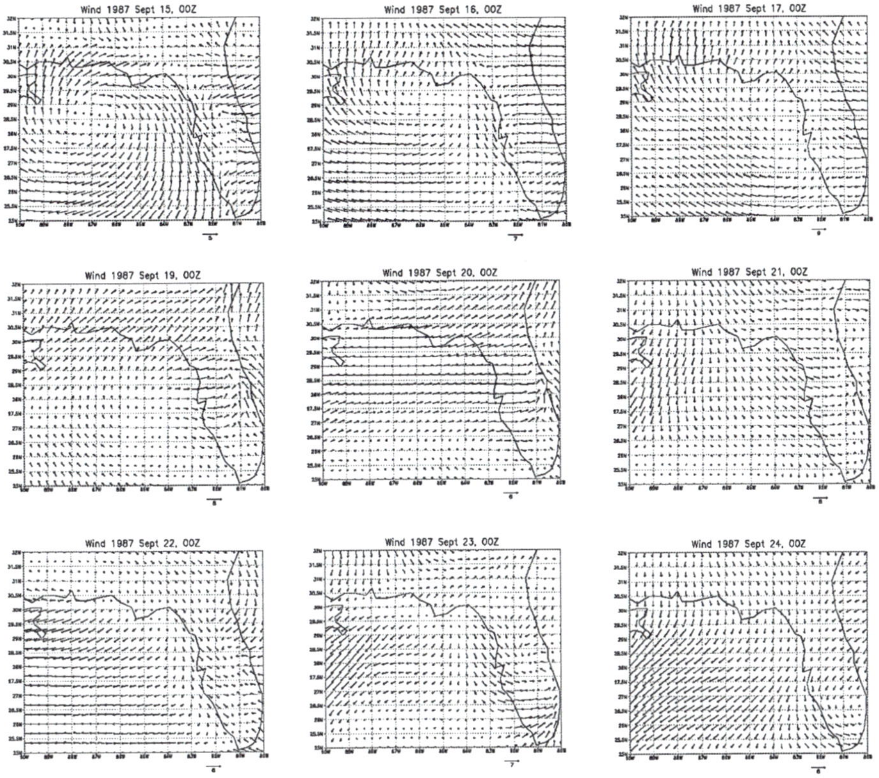

Figure 7.
NARR-based wind fields over the west coast of FL on August 24 1987, during the numerical passive tracer experiment.

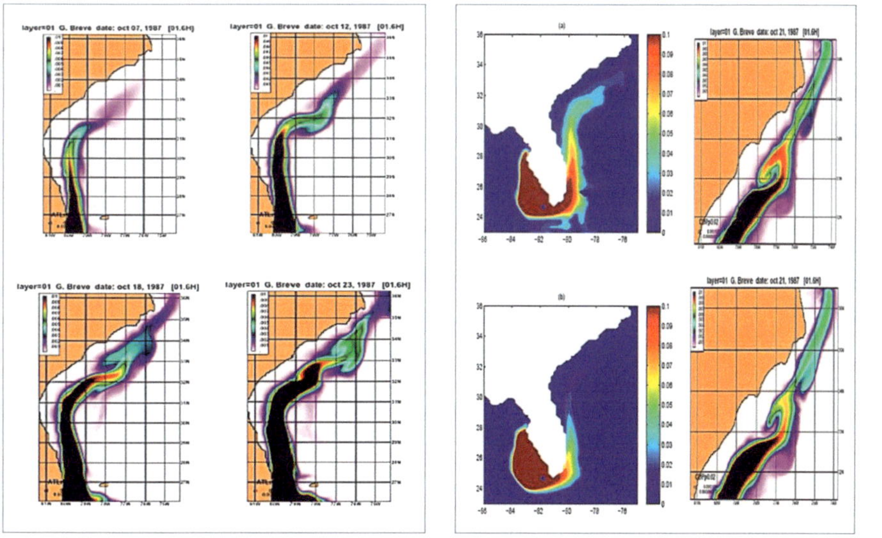

Figure 8.
The passive tracer wind-driven numerical model experiments employing the NARR versus the ECMWF driving fields, both driving the ROMS Ocean currents model. (a) Left panel, the passive K-breve algae; (b) right panel, a surface oil spill in the GOM being transported to the southeastern Atlantic seaboard.

6. The atmospheric wind field and extra-tropical cyclones

The wind field as observed by the National Weather Service (NWS) at the Cape Hatteras Lighthouse station (not shown) was evaluated over the period September 1 through December 31, 1987. The wind's velocity vector, i.e., wind speed and direction, is measured and recorded every 3 hours. The Cape Hatteras wind vector time series data were chosen since no meteorological buoy data were available (from the region) for the Fall of 1987. The Hatteras winds were deemed more representative of outer shelf Onslow Bay wind conditions than were winds from Wilmington or Beaufort, NC. Weisberg and Pietrafesa [14] found that in the Carolina Capes, wind speed increases from 1.5 to 2.5 times in magnitude along the coastal mainland to several tens of kilometers offshore due to the larger boundary layer drag created by land vs. that of water, which slips with the wind. The net result is that the effective wind stress over water is 2.5–6.5 times larger than that over land, albeit in the same direction. Cape Hatteras winds are less affected by the frictional boundary layer created by the mainland, because they are collected on a barrier island more than 20 kilometers from the mainland and are thus deemed more representative of actual over-the-water winds. It is of note that the region surrounding Cape Hatteras is a spawning region for wintertime atmospheric low-pressure systems or cyclonic storms [15].

During the late fall, winter, and early spring period, Atlantic low pressure systems known variously as Nor'easters, Atlantic Lows, Cape Hatteras Lows, and Extra-Tropical Cyclones (ETCs) are omnipresent over the coastal zone principally from South Carolina (SC) to Virginia (VA) [15] but actually extend from 25° N latitude to 75° N. ETCs intensify, and often form, throughout this zone, centered about Cape Hatteras [15]. The ETCs can deepen, i.e., further intensify, or spawn through a process known as "cyclogenesis" [16] and develop rapidly along and off the coast. The SC to VA coastal region is unique in its position adjacent to the warm waters of the Gulf Stream. Its alignment is favorable to the generation of offshore flow in response to winds typically associated with the incursion of cold, dry air from the north and west, often referred to as cold-air outbreaks (CAOs). The oceanographic setting in the region between SC and VA is such that the Gulf Stream Front (GSF) is omnipresent along the shelf-break between 32.5° and 35.5° N. During occasions of incursions of cold dry air streaming into the area from the north, local air temperatures can drop to between 0°C and 10°C, hence a CAO and the formation or genesis of an ETC. Cione et al. [15] determined that the mean path of the ETCs was from the SW to the NE and located about 30–50 km offshore so that the winds on the coastward side of the storm were from the NE to SW. As it occurs, the wind field present on the NC/SC coasts (**Figure 9**) was created by the passages of a series of ETCs.

The ETC winds would have driven offshore waters shoreward as depicted in **Figure 10** (from [3]). The basic dynamic balance relating the onshore-offshore component of the flow field in the surface layer of the water column is described by invoking conventional Ekman theory, in which the onshore-offshore (diabathic) mass flux, Mx, in the surface layer, D, is related to the alongshore (parabathic) component, Wy, of the total wind-stress vector or as expressed via: Mx = Wy/fD (1), where: Mx = Vertical Integral of r(udz) from 0 to D, the water surface to the depth D, the depth of the surface Ekman Layer, r is the water density, u is the diabathic or cross shelf water velocity, and f is the local Coriolis frequency. Note that this relationship states that the net transport of the wind-driven surface layer will be directly onshore if the wind is blowing from the northeast.

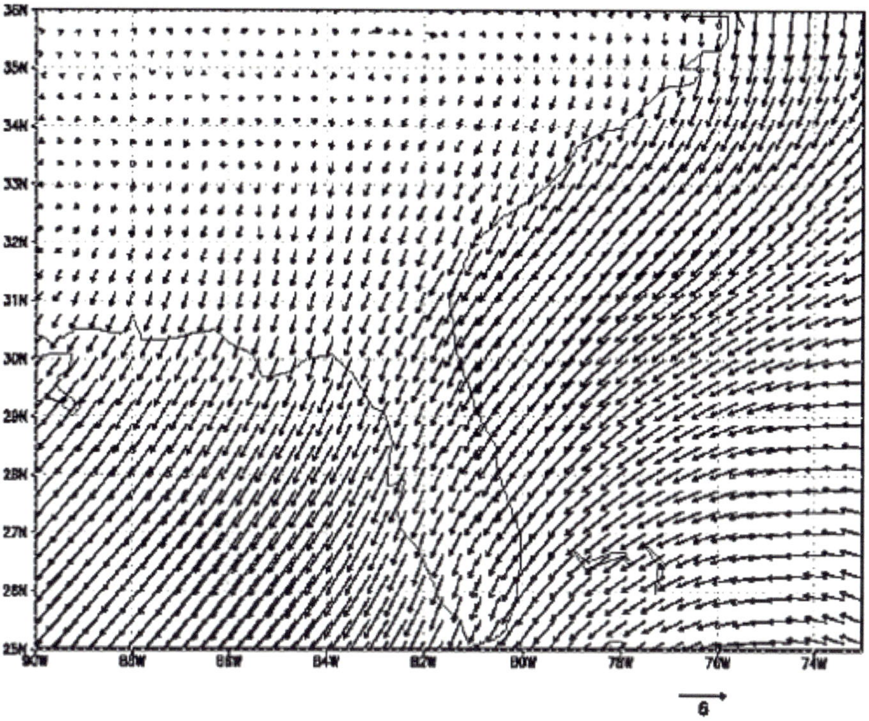

Figure 9.
The mean winds for the month of October 1987 from NARR winds (https://www.emc.ncep.noaa.gov›mmb›rreanl›index.html).

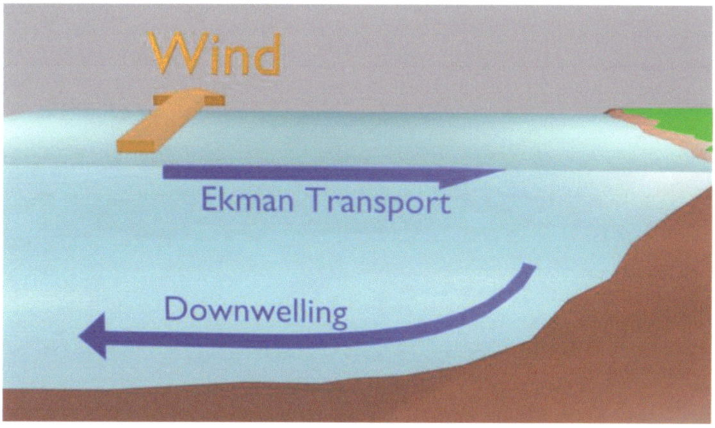

Figure 10.
Wind blowing with the coast to the right, creating a surface Ekman transport toward the NC and SC coasts.

The surface layer shown in **Figure 10** cartoon will be of the order of 5–25 m thick as a function of wind speed, vertical density gradient, and vertical velocity gradient on the NC and SC shelfs. Thus, a positive Wy (a northeastward wind, not shown) yields a positive Mx (surface layer transport to the southeast or offshore) and a negative Wy (a southwestward wind, as shown) yields a negative Mx (surface layer transport to the northwest or onshore). From October 9 through November 9, the wind velocity vector at Cape Hatteras was directed toward the southwest to south sector with essentially no reversals. From October 12 to 18, the winds were especially strong toward the south-southwest. On November 8th, the winds switched to become northeastward to northward. Over the entire 19-day period, October 9–27, the mean Wy, alongshore wind-stress component, was about 0.75 dynes/cm^2, which suggests a surface Ekman layer, D, of approximately 12–15 m thick and a vertically averaged onshore Ekman layer speed of approximately 6.3 cm/sec (or 5.5 km/day).

During the October 9–27 period, the distance water parcels and/or passive drifters would have moved across the shelf in the surface layer, which is about 105 km (65 mi). To calculate the total trajectory of a water parcel located in the surface layer requires that we integrate the vertically averaged (mean) onshore velocity component over the total time, with the wind fluctuating but remaining favorable for a shoreward moving surface water layer. For example, from October 13 to 16, the wind blew toward the SSW with an effective stress of between 1 and 3 dynes/cm^2, causing an onshore displacement of the surface layer of some 52 km (32 mi), about 13 km/day. By October 19, the surface layer had moved an additional 16 km (10 mi) shoreward driven by the SW winds of 0.3–0.5 dynes/cm^2. At this point, a passively drifting, buoyant particle imbedded in the GSF prior to October 8 would have traversed some 76 km (47 mi) across the shelf. To evaluate the possibility of this having occurred, we check the AVHRR imagery of October 19.

In **Figure 11**, the NOAA AVHRR SST map created by Dr. S. Baig (AOML) is shown. It appears that the entire Gulf Stream Front system of three filaments, which were present on October 9 (**Figure 5** lower panel), were subsequently mechanically driven

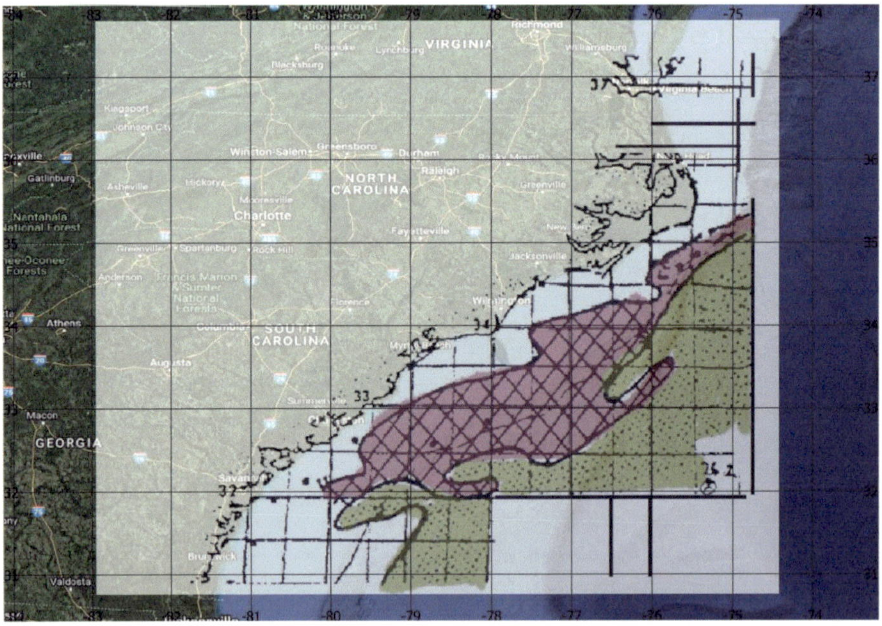

Figure 11.
NOAA AVHRR image of the Gulf Stream and GSF on October 19 1987 (courtesy of Dr. S. Baig, AOML). The magenta coloring is employed to depict the Gulf Stream surface waters that have been mechanically driven by the atmospheric winds towards the coasts of the Carolinas. The green coloring is employed to depict the Gulf Stream and its Frontal Filaments.

or rather, advected, onshore. The thermal frontal feature located in midshelf waters suggests that frontal waters, which 10 days previous were part of three filaments, now blanket the mid to outer shelf of Raleigh and Onslow Bay NC and Long Bay NC/SC. Amazingly, the warm-water front appears to have maintained its general outline, essentially intact, from 10 days earlier. From a comparison of **Figure 5** (lower panel) and 10, it is clear that the warm water boundary defining the filament front has moved more than 70 km across the shelf.

From October 17 to 27, the surface layer was advected another 26 km shoreward. By the latter date, the first warm water parcels that were mechanically detached from the Gulf Stream 19 days earlier would have reached the shoreline of mid Onslow Bay NC. By October 31, the entire Onslow Bay coastline could have been invaded by filament waters. Then, from October 31 to November 7, the southwestward winds would have blown an additional water mass 18 km wide in the onshore/offshore direction and 13 m thick toward the coast. In all, a block of water 100 km wide in the longshore direction, 13 m thick in the vertical, and 163 km wide in the cross-shelf direction moved across Onslow Bay coastal waters over the 30-day period. This is depicted in **Figure 11** as the rosette-colored water masses. So, every day, on the average, a block of water 40 feet thick, 62 miles long, and 3.3 miles wide was advected toward the coast. At least 12 of those blocks reached the NC and SC beaches. That scenario, depicted inferentially by winds and SST images, raises the question: Could this scenario be numerically modeled to validate the data-based explanation of the physical dynamics? We address that next.

An additional numerical model experiment, employing the National Weather Service Weather Research Forecast Model (WRF), described in Skamarock et al. [17] and ROMS. If oil were to be drilled in NC waters, and an oil spill occurred between September and March, when the winds are predominately out of the North to East Quadrant (Weisberg and Pietrafesa, 1983), then the oil would likely reach NC coast as shown in **Figure 12a**. If the winds were absent, but a GSF was passing by (e.g., **Figure 4b**), particularly where these filaments nearly hit the Outer Banks near Cape Hatteras NC, then the oil spill would be carried as shown in **Figure 12b**. SC has a ban on oil drilling, so it was not considered in the latter two experiments.

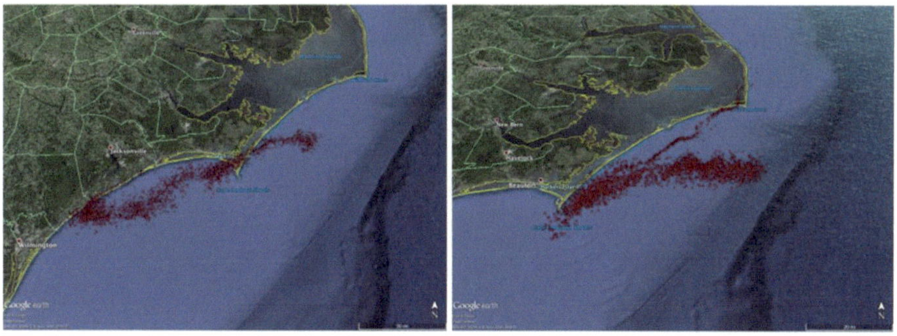

Figure 12.
The numerically modeled trajectory of surface oil spills in central Raleigh Bay NC: (a) left panel, shows the surface trajectory of a virtual oil spill driven by an ETC from the middle of the northern bay to the southwest into the lower bay; (b) right panel, shows the surface trajectory of a virtual oil spill offshore and then entrained into a passing GSF. Both wind driven (a) and ocean feature (b) events, project oil being carried to the NC coast. The Red Dots are employed in both panels to depict the trajectories of the oil spilled offshore and carried via surface currents.

7. Conclusions and summary

Oil Spills have never occurred on the North Carolina or South Carolina US continental margins. However, they have occurred in the Gulf of Mexico where a great many oil drilling platforms are located, and many oil tankers transit Gulf Waters daily. In 2010, there was a major oil spill in the Gulf. In 2017, the US administration lifted a ban on oil drilling in North Carolina Coastal waters. So two questions arise: One, could an oil spill in the Gulf reach Carolina coastal waters? Secondly, if an oil spill were to occur in the future off North Carolina, could it reach the beaches under typical environmental conditions? We address both questions.

To address non-local oil spills reaching the Carolinas, we revisit a Red Tide outbreak on the West Florida Shelf that reached the coasts of the Carolinas. In August 1987, a Red Tide occurred on the West Florida continental shelf off Naples, Florida. Those deleterious plants reached North and South Carolina shelf waters by late October. Using data from that era, we create a data-based scenario by which atmospheric conditions combined with oceanic currents carried the Red Tide plants over 1600 kilometers (1010 miles). Then based on the surrogate Red Tide plants, we numerically modeled the atmospheric and oceanic conditions of 1987 and dumped passive oil particles into the numerical model off Naples, Florida. The model predicted that the oil would have reached Carolinas coastal waters. Thus, we demonstrate from both data and numerical modeling that oil spills in the Gulf of Mexico have the potential to reach the US eastern seaboard via a combination of atmospheric storms, major ocean currents, and atmospheric wind-driven surface currents.

To address the possibility of a local oil spill reaching the beaches of the Carolinas, we consider the 2017 U.S. White House removal of an oil-drilling ban in North Carolina coastal waters. Herein, we released oil spilled in a projected oil drilling location on the North Carolina shelf into our atmospheric and oceanic numerical model. Again, given typical atmospheric and oceanic phenomena, the hypothetically spilled oil reached the beaches of North Carolina.

Thus, we have demonstrated that as a matter of a series of non-local and local atmospheric and oceanic phenomenological consequences, in outer continental shelf waters, oil in the upper 15 meters of the water column can be carried long distances and then driven toward the coast and reach the beaches. This is especially true during the passage of an atmospheric Extra-Tropical Cyclone. If a Gulf Stream Frontal Filament is present offshore, then during the passage of a cyclone event, oil in the upper 15 meters, even well offshore (~75 m deep) will be carried to the coast.

Acknowledgements

The authors acknowledge the following agencies and organizations for support of this study: The National Oceanic & Atmospheric Administration—National Marine Fisheries Service, the Southern Environmental Law Center, the Bureau of Ocean Environmental Management, and the Z. Smith Reynolds Foundation.

Author details

Leonard J. Pietrafesa[1,2]*, Shaowu Bao[2], Paul T. Gayes[2] and Farid Askari[2]

1 Department of Marine, Earth and Atmospheric Sciences, North Carolina State University, Raleigh, North Carolina, United States

2 Center for Marine and Wetland Studies, Coastal Carolina University, Conway, South Carolina, United States

*Address all correspondence to: ljpietra@ncsu.edu

References

[1] Pietrafesa LJ, Janowitz GS, Askari F. The red tide invasion of Carolina coastal waters. NOAA UNC-SG pub. WP-88-01, Grant No. 85AA-D5G022, 36 pp. 1988

[2] Pietrafesa LJ, Bao S, Gayes PT, Carpenter D, Kowal J. On the variability and trend of the Florida current. Journal of Coastal Research. 2022 (in press)

[3] Pietrafesa LJ, Janowitz GS. On the dynamics of the Gulf stream front in the Carolina Capes. In: Proceedings of the Second International Symposium on Stratified Fluids. Trondheim Norway. Physical Oceanographic Processes in the Carolina Capes. 1980. pp. 184-197

[4] Bao S, Pietrafesa LJ, Churchill JJ, Jacobs N, Gayes PT. Observations and Modeling of Cyclogenesis off Cape Hatteras North Carolina. Advances in Meteorology; 2022

[5] Pietrafesa LJ, Blanton JO, Wang JD, Kourafalou VH, Lee TN, Bush KA. The tidal regime in the South Atlantic bight. In: Atkinson LP, Menzel DW, Bush KA, editors. Oceanography of the Southeastern US Continental Shelf. Washington, DC: American Geophysical Union; 1985. pp. 63-76

[6] Pietrafesa LJ. Evidence of the deflection of the Gulf stream at the site of the Charleston bump. The Gulf Stream. 1975;**4**:4

[7] Rooney DR, Janowitz GS, Pietrafesa LJ. A simple model of the deflection of the Gulf stream by the Charleston rise (bump). The Gulf Stream. 1978;**4**(11):3-7

[8] Pietrafesa LJ. Continental shelf oceanographic and atmospheric processes in the Southeast U.S. In: Barnett W, Riggs S, editors. Phosphorus Deposits of Continental Margins. American Geophysical Union Press; 1988. pp. 1-32

[9] Sun CS, Pietrafesa LJ. Gulf Stream Meander Waves, a Numerical Model. Journal of Non-Linear Mathematics. Baroclinic-Barotropic Instabilities and Energy Transfers in the Gulf Stream. In: Proc. First World Congress of Nonlinear Analysts. Tampa, FL: International Association of Nonlinear Analysts; 1992. pp. 3515-3532

[10] Johns WE, Schott F. Meandering and transport variations of the Florida current. Journal of Physical Oceanography. 1987;**17**(8):1128-1147

[11] Pietrafesa LJ, Janowitz GS. A note on the identification of a Gulf stream spin-off eddy from Eulerian data. Geophysical Research Letters. 1979;**6**:549-952

[12] Pietrafesa LJ. Survey of a Gulf stream frontal filament. Geophysical Research Letters. 1983;**10**(3):203-206

[13] Shchepetkin AF, McWilliams JC. The regional oceanic modeling system (ROMS): A split-explicit, free-surface, topography-following-coordinate oceanic model. Ocean Modelling. 2005;**9**(4):347-404. DOI: 10.1016/j.ocemod.2004.08.002

[14] Weisberg, RH, Pietrafesa LJ. Kinematics and correlations of surface winds in the South Atlantic bight. Journal of Geophysical Research, Oceans. 30 May 1983;**88**(C8);4593-4610

[15] Cione JJ, Raman S, Pietrafesa LJ. The effect of Gulf Stream-induced baroclinicity on the U.S. east coast winter cyclones. Monthly Weather Review. 1993;**121**:421-430

[16] Bosart LF, Vaudo CJ, Helsdon HJ Jr. Coastal frontgenesis. Journal of Applied Meteorology. 1972;**11**:1236-1258

[17] Skamarock WC, Klemp JB, Dudhia J, Gill DO, Barker DM, Duda MG, et al. A Description of the Advanced Research WRF Version 3. NCAR Technical Note NCAR/TN-475+STR. 2008. DOI: 10.5065/D68S4MVH

Chapter 5

Computational Techniques of Oil Spill Detection in Synthetic Aperture Radar Data: Review Cases

Singanamalla Vijayakumar

Abstract

In this chapter, a major role of environmental assessment is an oil spill identifies or detected from the coastal region surfaces or marine surroundings. Normally, the oil spills on the coastal regions impact their characteristics of environmental activities. However, these activities are monitoring through several radar satellites and sensor. For those achievable activities detecting or identifying, many researchers developed several approaches. Particularly, this chapter discusses about the detection of oil spill current operational effects on coastal region surfaces. In addition, the current research operations of oil spill characterizations and quality of its impacts, effects of current environmental bio-systems, their control measurement strategies, and its surveillance operations are discussed. Finally, the oil spill detection is done through the SAR image region classification based on its feature extraction. This could be monitored from the image dark region selection through remote sensing techniques.

Keywords: remote sensing, oil spill detection, coastal monitoring, pollution control, SAR images

1. Introduction

The coastal region monitoring is very important to protect marine environment from oil spill pollution, which is caused due to oil transportation, storage, tanker accident, or intentional drainage of oil into sea water. Day by day the percentage of pollution-affected coastal or ocean region due to fuel is getting increased and thus, it becomes increasingly important to protect ocean regions and lives of marine regions [1, 2]. The spilled oil of ships or tankers spread over the sea surface forms dark regions and causes marine pollution. It could affect the lives of the coastal regions to greater extent. Thus, a suitable mechanism is required to detect and remove those oil spills from coastal regions at the earliest. The above-said task could be accomplished only through remote sensing technique, which is defined as a coherent technique through which any regions or objects in the earth surface could be located at a far distance and monitored. In remote sensing process, information about earth's object is collected either using space-borne satellite or using airborne aircraft [3]. At the initial stage, space-borne satellites could be used for locating the region but further analysis would be carried out with the help of airborne aircrafts. In order to detect and remove spilled

IntechOpen

oils on sea surface, space-borne satellites such as ERS, ENVISAT, RADARSAT are predominantly used. In general, there two kinds of RADAR are in use, they are Synthetic Aperture Radar (SAR) and Real Aperture Radar (RAR). The Synthetic Aperture Radar (SAR) uses microwave sensors for data collection in the form of 2-D images [4, 5]. These Synthetic Aperture Radars are mounted on satellites and used for earth observations. The microwave sensors reflect the image visual perception of any earth objects and their attributes. Thus, it could be used for monitoring ocean regions during day and night under all weather conditions for detection of spilled oil. Norwegian water pollution authorities are using space-borne satellites for monitoring oil pollution [6].

In this chapter, extensive study on oil spill detection using both manual and semi-automated approaches is carried and presented. The objective of this study is to determine various approaches currently being used for oil spill detection and its accuracy in ocean monitoring for environmental protection. Though the research has been started in 1994, the automatic detection of oil spill is still in infant stage as it is considered as a very complex task. Since oil spill on sea surface resembles look-alike, it is very difficult to differentiate them automatically from it. In general, look-alikes of oil spill are biological film, grease ice, front eddies, rain cells, internal waves, and upwelling zones [7, 8]. In most of the cases, even the trained operator cannot differentiate oil spill from look-alikes, for example, it has been carried out manually at TSS by assigning attributes such as low, medium, and high value to various oil slicks based on its existence with attributes. Once it is characterized as oil spill, this information could be further analyzed through aircraft surveillance system used by Norwegian Pollution Control Authority (NPCA) [9, 10]. Thus, it is very difficult to design an automatic oil spill detection system.

In Section 2, the problems of oil spills in coastal regions were discussed. In Section 3, characteristics of Spilled oil and its impacts in coastal regions are presented in detail. Section 4 discusses about various sensors and satellites employed for ocean monitoring. Section 5 presents existing methodologies till date and Section 6 presents various classifications approaches using features of Oil Slicks. In Section 7, current operational efforts on oil spill monitoring is discussed elaborately. Section 8 briefs case studies in relevant area. Section 9 concludes the work.

2. Oil spill problems

In this section, the major problem in coastal region monitoring is oil spill detection. The spills are more predominantly considered during marine ecosystem pollution control. However, the marine oil spill term considers—where oil is released into costal or ocean waters and spills may also occur on land.

From these circumstances, the oil spills are very harmful to marine birds, sea turtles, and mammals, and also can harm fish and shellfish. In general, the oil spills are destroying the insulating ability of fur-bearing mammals, such as sea otters, and the water-repelling abilities of a bird's feathers, exposing them to the harsh elements. Many birds and animals also swallow oil and are poisoned when they try to clean themselves or when eating oiled prey [11].

This can be considered to monitor and detect the oils pills on coastal surface areas. To protect the coastal environments, this could be considered to monitor the ocean regions and also surfaces of oceans.

Finally, many techniques are reviewed to carry such type of problems in marine regions. All these techniques/methodologies/approaches are discussed in subsequent sections.

3. Characteristics of spilled oil and its impacts

In this section, characteristics of spilled oil on coastal region and its impacts are discussed in detail. The spilled oil can spread over sea water, on sea shore, land area, and glaciers. It is highly dangerous as wind, waves, and currents could scatter large oil spill over a wide area within a short period of time [12]. The spilled oil quickly spreads to form a thin layer on the water surface, which is called as oil slick. As time progresses, the oil slick becomes thinner, forming a layer called as a "sheen," which has a rainbow-like appearance with a multiple color. The light weigh oil is highly toxic but it gets evaporated quickly, whereas heavy weigh oil is less toxic but persists in the environment for a longer time. The heavy weigh oil can get mixed with pebbles and sandy beaches where they may remain for many years [13].

It is required to measure the spread of spilled oil over the sea surface called "slick measurement." It is achieved through the wavelength of backscattering of capillary waves that are generated only on the very dark region of sea surface. The measurement of slick regions also depends upon the polarization of RADARs and its incidence angle, nature of spilled oil, and metrological ocean conditions [14].

The coastal region monitoring depends on the capability of radar satellite sensors to be used to detect the oil spill on the sea surface. The multi-temporal imaging technology is used to obtain data captured by sensors as it is very important to model the oil spill. Satellite remote sensing sensors can provide the following information for oil spill contingency planning of large area location and oil spill spread, thickness of the oil spill quantity estimation, the classification of the oil spills from the environmental activities and finally assist the information that provides a cleanup valuable operation [15].

The oil spill in water may severely affect the marine environment and causes damage to plankton and other aquatic organisms. The livelihood of many coastal people is severely impacted by oil spill, particularly those who are depending upon fishing and tourism as their livelihood [16].

The movement of oil on land surface depends on various factors such as oil type, soil type, and moisture content of the soil. Oil spilled on agricultural land can impact on soil fertility and pollute ground water resources as well. Oil companies and shipping operators are responsible for controlling spilled oil and cleaning polluted areas. In the event of oil spill, information about the size and extent of it is very critical and it is required to assist the government and industry in oil spill contingency planning to describe the guidelines for estimating oil thickness using visual appearance as shown in **Table 1** [17]. It also explains the appearance of oil varies from silvery-sheen to dark brown with its approximate thickness in micrometer.

Oil spill appearance	Approximate film thickness (μm)
Silver sheen	0.05
Rainbow sheen	0.15
Reddish-brown sheen	0.50
Brownish	2.00
Dark	10.00
Dark Brown	50.00

Table 1.
Appearance of oil on water surface.

3.1 Impacts of spilled oil

In **Table 2** nature of oil spill, environment, and its impacts are presented in detail [18]. This is clearly explained about the present impacts of oil spills in environments. Each step gives the nature of present impact of spilled oil in two types of regions: one is called, Aqueous Region and another called, Land Surface. The first, Aqueous Region explains an impact of the current spilled oil on the ocean water surface and its affecting on the environments from the chemical organs. Second, Land Surface gives the current role of spilled oil about the groundwater bodies and its effects on earth surface nature.

However, in the physical and bio-degradable components of environmental persistence are accumulative processes on the normality for minimizing the oil parameters

Nature of spilled oil	Spilled environment	Impact of spilled oil
Lighter oil less dense than water	Aqueous region	Spilled oil spreads over surface of sea water and it affects the livelihood of aquatic organisms
	Land surface	Spilled oil penetrates the earth surface and moves down due to gravity and reaches groundwater bodies and contaminates it.
Mixture of volatile compounds (hydrocarbons)	Aqueous region	Spilled oil started vanishing when in contact with air and pollutes it. It happens once it reaches top point of water surface.
	Land surface	Volatilization may happen likewise from a spill amassed on top of shallow groundwater. Once vanished, the unstable vaporous mixes with air and pollute environment.
Mixture of compounds with different water solubility	Aqueous region	In spite of the fact that oil is little water solvent and collects on top of water, some oil mixes may break down to some degree in water. The broke down mixes turn out to be more portable and spread bit speedier with water and more bioavailable to be taken by marine life.
	Land surface	Oil is less leachable by precipitation water and may continue in subsurface situations, including on top of groundwater for a considerable length of time.
Sheens or Oil Slicks	Aqueous region	Since oil and water are not miscible, it forms shapes on water surface. This expands the resistance of oil to characteristic constriction procedures and makes it steadier in the earth.
	Land surface	Oil is less leachable by precipitation water and may continue in subsurface situations, including on top of groundwater for quite a long time.

Nature of spilled oil	Spilled environment	Impact of spilled oil
Oil physical state (liquid) and biodegradable components	Aqueous region	Some normal constriction procedures may lessen the measure of dirtied oil even without human intercession. This is because of a blend of vanishing, disintegration, and biodegradation of oil slick. In marine or stream situations, such characteristic lessening procedures are quicker than those in subsurface situations, ashore or in groundwater.
	Land surface	
Oil composition involving the presence of individual components with environmental persistence and bio-accumulative potentials	Aqueous region	Some oil segments (e.g., PAHs) may continue longer in environment, collecting in dregs, marine life, fishes, and natural life when all is said in done.
	Land surface	

Table 2.
Impacts of spilled oil in environment.

without the human interaction. Because of the coastal condition, the marine environment properties are minimizing the procedures as well as surface situations or ground water conditions.

3.1.1 Impact of freshwater habits

Oil spills occurring in freshwater bodies are less publicized than spills into the ocean even though freshwater oil spills are more frequent and often more destructive to the environment. Freshwater bodies are important for human health and the environment, but unfortunately, it is highly sensitive to oil spills [19]. They are often used for drinking water and frequently serve as nesting grounds and food sources for various freshwater organisms. All types of freshwater organisms are susceptible to the deadly effects of spilled oil, including mammals, aquatic birds, fish, insects, microorganisms, and vegetation. In addition, the effects of spilled oil on freshwater microorganisms, invertebrates, and algae tend to move up the food chain and affect other species [20].

3.1.2 Impact of marine environmental bio-system

The major spills of crude oil and its products in the sea occur during their transport by oil tankers, loading and unloading operations, blowouts, etc. Due to spillage of oil in the marine environment, it undergoes a variety of transformation involving physical, chemical, and biological processes. The physical and chemical processes include evaporation, spreading, emulsification, dissolution, sea-air exchange, and sedimentation. Chemical oxidation of some of the components of petroleum is also induced in the presence of sunlight. The degraded products of these processes include floating tar lumps, dissolved and particulate hydrocarbon materials in the water column, and materials deposited on the bed [21].

Biological processes, though very slow, also act simultaneously with physical and chemical processes. The important biological processes include degradation by microorganisms to carbon dioxide or organic material in intermediate oxidation stages, uptake by large organisms, and subsequent metabolism, storage, and discharge [22].

3.1.3 Impact on marine habitats

The marine environment is made up of complex inter-relations between plant and animal species and their physical environment. Harm to the physical environment will often lead to harm for one or more species in a food chain, which may lead to damage for other species further up the chain. Where an organism spends most of its time—in open water, near coastal areas, or on the shoreline—will determine the effects of an oil spill is likely to have on that organism [23].

In open water, marine organisms such as fish and whales have the ability to swim away from a spill by going deeper in the water or further out to sea, reducing the likelihood that they will be harmed by even a major spill [24]. Marine animals that generally live closer to shore, such as turtles, seals, and dolphins, risk contamination by oil that washes onto beaches or by consuming oil-contaminated prey. In shallow waters, oil may harm sea grasses and kelp beds that are used for food, shelter, and nesting sites by many different species.

4. Sensors and satellites

In coastal region monitoring, oil spill detection and its removal play an important role in protecting the environment and reducing economic loss. It could be achieved through remote sensing technique that uses space-borne satellites and airborne aircrafts. The space-borne satellites use mounted RADAR in it which comprises sensors. It is also observed that the selection of particular sensor plays an important role in detecting oil spills. At the initial level, space-borne satellites are most suitable for oil spill detection but further analysis could be carried out using aircraft sensors [6]. In ocean monitoring, microwave sensors are preferred to optical sensors due to its excellent performance under all weather conditions. The role of different sensors in oil spill detection is presented in the subsequent section.

At the initial stage oil slicks have been identified and processed using images of ERS, ENVISAT ASAR, and RADARSAT-1. Recently, many radar satellites with a potential for oil slick location have been propelled, specifically, RADARSAT-2, Terra-SAR-X, and COSMO SkyMed. A general exchange of the suitability of RADARSAT-2 for beach-front applications is found in [17]. These satellite radars use new imaging modes that open some new potential outcomes for enhanced imaging of oil slicks.

However, the ERS-1 is currently being used by the Norwegian Space Center (NSC) on behalf of pollution control board in the coastal region for oil spill detection. In ERS-1, radar frequency range is located accurately and it is used for processing radar signal-by-signal projections. It is the technique used to capture images of resolution 30 m X 30 m. In ERS-1 SAR raw data processing, large geographical area can be covered by adjusting physical length of the ERS-1radar antenna. In specific, for oil spill detection, the ERS-1 radar produces variety of images at various levels based on wind speed by covering large area. These images need to be processed to identify oil slicks. At present, ENVISAT RADAR is predominantly used for oil spill detection due to its wide area coverage with a spatial resolution of 150 m and a pixel spacing of 75 m. Similarly, the RADARSAT, ScanSAR covers 60–400 km swath width with a spatial determination of 50 m and a pixel dispersing of 25 m [25].

Even though the ENVISAT higher-frequency radar, spatial data determination is accessible from the remote sensing medium determination, the spatial data information takes care a sensible bargain permitting covering the vast territories and as yet recognizing significantly littler oil slicks from the SAR image regions [26].

The other is named as COSMO-SkyMed satellite with a group four, which permits the great scope of feature analysis in significant time periods. The COSMO-SkyMed SAR fragment was dispatched in November 5th, 2010 at 1920 L and it has been utilized for the water/deep-water horizon oil slick studies. As indicated in [27], it ought to have greater capability in detecting oil slick localities with double polarimetry SAR images that are captured by COSMO-SkyMed satellite sensors.

Terra-SAR-X has the capability of capturing images in different directions while monitoring the specified area of coastal region. It uses larger value of standard deviation for differentiating oil slick or look-alike. In addition, it uses double polarimetric SAR technique, in discriminating the spilled oil from look-alike. The RADARSAT-2 has a few polarization alternatives—for example, single-polarization SAR modes with one of the accompanying polarizations such as HH or HV or VH or VV [28]. It also offers dual polarization SAR modes with appropriate polarizations such as HH + HV or VV + VH. A quad-pol alternative is likewise feasible for fine and standard modes.

Adamo et al. have proposed a model for oil spill examination by considering normal number of accessible MODIS, MERIS, and ASAR images. These images need to be captured within observing time periods without cloud images in sunglint conditions. However, the MODIS/MERIS is ensured if the smoothness of radar data as image is situated in oceanography regions inside as well as outside. Similarly, they noticed a wavelength from reliance oil/non-oil class distinctness with expanded execution at

Satellite name	Agency	Sensor name	Spatial resolution	Spectral resolution	Year of launch
RORSAT	NOSS	EORSAT	Low	X-band	1967
Seasat	NASA	SMMR	High	L-band HH polarization	1978
GeoSat	ESA	GRA	Medium	Ku-band	1985
ERS-1	ESA	GOME SAR	Low	C-band VV or Sun-Syn pol	1991
ERS-2	ESA	GOME, ATSR-2	High	C-band VV or Sun-Syn pol	1995
RADARSAT-1	CSA	SCN	Medium	C-band HH	1995
GFO	ESA	GFO-RA	High	Ku-band	1998
JERS-1	JAEA/NASDA	OPS	High	L-band HH polarization	1998
Quickscat	NASA	SeaWinds	High	Ku-band	1999
ALOS	NASDA	AVNIR-2	Low	L-band HH	2001
ENVISAT	ESA	MODIS, MERIS (P) and ASAR WSM	Low/ medium/ high	7 bands, Q 15 bands, P and C-band VV or HH	2002
ALOS	NASDA	PALSAR	Fine resolution	L-band HH polarization	2006

Satellite name	Agency	Sensor name	Spatial resolution	Spectral resolution	Year of launch
SAR Lupe	ESA	XSAR	High	X-band	2006
Metop – A	ESA	ASCAT	High	C-band	2006
RADARSAT-2	CSA	SCN/QP	Medium/ high	C-band HH or VV or Quad pol	2007
TerraSAR-X	CSA	TSX-SAR	High	X-band Single or dual pol	2007
COSMO-Skymed	ISA	Multi-SAR	High	X-band Single or dual pol	2007
Cloudsat	SDSC	CPR	High	X-band	2008
RISAT-2	SDSC (ISRO&IAI)	XSAR	High	X-band	2009
TanDEM-X	ESA	GSCDA	High	X-band	2010
RISAT-1	SDSC (ISRO)	ScanSAR	Medium	C-band	2012
Metop – B	ESA	IASI	High	X-band	2012

Note–P: 350–1050 nm, Q: 610–2155 nm.

Table 3.
Different space-borne satellites with its specifications for remote sensing.

the more NIR band wavelengths of MODIS and MERIS [27–29]. Their recreations for a range in the Mediterranean appeared that by and large 150 images every year would be suitable if an aggregate cloud list of under 2/10 was required. Occasional varieties further demonstrated that the period between spring and pre-winter was the most suitable for sunglint conditions. MERIS and MODIS symbolism is routinely used to make water quality maps and checking algal sprouts [30]. Indeed, without sunglint conditions, algal blossom maps can be utilized in the mix with SAR pictures to lessen the quantity of false cautions because of green growth. Radar satellites are most suitable for quick detection and monitoring of large areas. At present, most of the remote sensors can detect oil spills.

Table 3 shows various satellite-based sensors for oil spill monitoring. From these satellites, sensors are using various spectral bands (i.e., X, L, C, and Ku-band) for monitoring the coastal region surfaces. Finally, the remote sensing manner transmits the data from one location to another location, while observing the ocean surfaces for change detection of costal activities.

Band type	Frequency range (GHz)	Wavelength range (cm)
C-band	4–8	60–30
L-band	0.5–1.5	30–15
Ku-band	10.7–12.75	2.5–1.67
X-band	8–12	3.75–2.5

Table 4.
Different band and its wavelength with its frequency range.

In **Table 4**, different bands used by different satellites are shown below with its frequency range and its wavelength. It can be considered as for the satellite band type as well as the frequency and wavelength ranges. The corresponding radar satellite only uses these appropriate band models. And also, most of the radar satellites are using a C or X-band for the coastal region monitoring.

5. Existing methodologies

A review of literature survey on various methodologies used to detect oil spill for ocean monitoring has been carried out and presented subsequently in detail.

In 1999, Solberg et al. [31] have proposed an algorithm for automatic detection of oil spill using rule-based approach combined with statistical modeling (i.e., Gaussian density). It also uses prior knowledge about oil slick for classification of it. In this algorithm, authors have used 11 features of oil slick that include distance to point source, number of detected spots in the scene, number of neighboring spots, Homogeneity, Slick Complexity, Slick width, slick area, first invariant planar, local area contrast, border gradient, power-to-mean ratio. The accuracy attained by rule-based classification approach is 95%. But the observed limitation of this approach is it requires prior knowledge about oil slicks. The proposed algorithm could be considered as semi-automatic algorithm as it uses prior knowledge.

In 2000, Fabio Del Frate et al. have proposed a neural network algorithm for classification of oil spill from look-alike in [1]. The proposed algorithm uses the rule-based Bayesian statistical decision approach. The performance of classification algorithm has been evaluated on a dataset containing oil spill and look-alike. The proposed algorithm uses some of the features that describe both physical and geometrical properties of oil spill. Those features are used as candidate features for classification. The accuracy attained by proposed algorithm is 70%. However, the observed limitation of this approach is that it is required to develop the classification rules for the complex datasets. Hence, this algorithm could be considered as neural network algorithm as it uses statistical-based classification decisions.

In 2004, Maged Marghany [32] has proposed an approach for oil slick detection and oil slick trajectory model. The proposed approach uses two sub-models. The first model uses texture analysis for oil slick detection. The second model uses the oil slick trajectory forecasting model. The oil slick trajectory model contains the integration between Doppler frequency shift model and Lagrangian model. Both models have used a classical Fay's algorithm, to simulate the oil slick trajectory movement. In this proposed algorithm, authors have used features such as entropy, surface current changes, energy, slick homogeneity, and slick direction. The results of this approach are not accurate, which is a primary limitation. The proposed approach could be considered as an automatic algorithm as it uses slick trajectory models of classical Fay's algorithm.

In 2004, Frederic Galland et al. [33] have proposed an approach for monitoring pollutant of major environmental hazards at ocean surface and minimizing it. The proposed approach consists of two parts. In first part of the proposed algorithm, homogeneous regions are partitioned that should be applied with polygonal active grid. In second part of the proposed algorithm, classification is carried out and it should be applied with an automatic minimum description length (MDL) thresholding technique. In this work, feature considered includes size, shape, distance of object from coastal region, intensity, and texture. The results of accuracy of oil spill regions are obtained for larger areas with appropriate coastlines. In this proposed method, observed limitation of MDL application is that noise is not removed fully from SAR images.

In 2005, Iphigenia et al. [34] have proposed computational intelligence-based approach for oil spill detection on satellite images. In this approach, smoothing is performed through Gaussian filtering followed by thresholding is applied. The image enhancement technique is applied to enhance dark regions and then, objects are grouped together using segmentation method. Finally, fuzzy classification approach is applied to classify oil spill or look-alike. In this approach, authors have used three features: region eccentricity, land distance and the probability of oil spill. The proposed algorithm has been tested with 26 images and attained 88% of accuracy in detecting oil spill regions.

In 2005, Fanny Girard-Ardhuin et al. [35] have proposed semi-automatic approach for detecting spilled oil regions, characterization of region properties, and classification of oil spill or look-alike. For semi-automatic detection of dark regions median filtering is applied to remove noise followed by Sobel operator for thresholding and segmentation. The filtered regions are characterized using features such as size and shape. In this algorithm authors have used some of the features about the oil spill formation of wavelength, spill polarization of sea surface areas, satellite projection of incidence angles with radar projection view, the slick nature and changes of the sea surface conditions. While applying this algorithm, the image dark region parameters are not properly detected and its complexity is not determined accurately.

In 2006, Gregoire Mercier et al. [36] have proposed a semi-supervised approach for oil spill detection using wavelet decomposition. This proposed method is implemented in complex sea surface and features used are sea surface wavelength, spill polarization of image region, the radar different views of oil spill projection. The obtained results of oil spill are shown as very effective and accurate in detecting oil slicks from the specified image region. This proposed approach is applicable for large ENVISAT images with complex formatting. It is considered as limitation of this algorithm.

In 2007, Topouzlis et al. [37] have proposed a neural network-based approach for detecting and classifying oil spill. The proposed approach detects the dark formation areas and classifies it into oil spill or look-alike. In this proposed algorithm, features such as oil spill perimeter, shape factor of the region, the object complexity estimation, standard deviation, and power-to-mean ratio are used. The obtained results show that the algorithm detects dark formation with 94% accuracy and discriminate it as oil spill with 89% accuracy. The limitation of this proposed algorithm is that it is not suitable for high dynamic data environments.

In 2007, Solberg et al. [38] and Maurizio Migliaccio et al. [39] have proposed a three-step algorithm for detection of oil spill using the RADARSAT and ENVISAT SAR images. The first step is used to detect the dark spot and the second step is used to extract the dark spot features, and finally, it classifies the dark spot as oil spill or look-alike. In this proposed algorithm, authors used features such as the complexity, width, area, and slick moment; contrast of dark region about local contrast, border gradient and smoothness of the spill region; homogeneity of spill region about the power-to-mean ratio; the spill surroundings about the detection of dark spots, spot region, and distance from the coastline. The result of this algorithm is reported for 59 images of RADARSAT and ENVISAT about oil spill or look-alike. The observed limitation is classification approach not carried automatically.

In 2008, Lena Chang et al. [40] have proposed an approach for ocean monitoring with minimized computational complexity. In this proposal, image segmentation is carried out using split and merge technique. The conventional detection theory called centralized look-alikes ratio test (GLRT) is used to detect oil spill using simple variance and statistical properties with the help of decision-making rules. The proposed algorithm determines oil spill effectively on ERS-2 SAR images. The similar pixels are not shown on the specified image region, while applying the GLRT method using ERS-2 SAR images.

In 2008, Camilla Brekke et al. [41] have proposed improved classification algorithm using a multivariate Gaussian classifier for reducing low-confidence levels. In this work, authors have used features such as slick complexity, power-to-mean ratio of slick, slick border, slick local contrast, slick width, slick region, slick smoothing contrast estimation, and slick variance. The observed limitation of this algorithm is that the manual operation could not be performed on selected region. In Ivanov et al.'s work [42] image processing methods are proposed for detecting oil spill. In this proposal, study is carried out on East China Sea for oil spill detection using SAR images. Authors have selected features such as shape, slick size or length, location, orientation, type of edge on oil slick, and database contrast. The result and analysis show small oil patches are uniformly distributed along the ship tracking. From the results, detection of oil spills on image region is not accurate. Kontantinos N. Topouzelis [43] have proposed a classification with fuzzy logic technique. In this proposal, authors have used nearly 25 features of oil spill includes object's area, perimeter, perimeter-to-area ratio, complexity, shape factor I, shape factor II, mean value, standard deviation, power-to-mean ratio, background mean value, background standard deviation, background power-to-mean ratio, ratio of the power to mean ratios, mean contrast, max contrast, mean contrast ratio, standard deviation contrast ratio, local area contrast ratio, mean border gradient, standard deviation border gradient, max border gradient, mean difference to neighbors, spectral texture, shape texture, and mean Horlicks texture. Authors have claimed that the obtained accuracy is reported as 99% in detection of oil spill. But this algorithm observed some limitations that are the imbalanced training datasets, data selection validity, and high dynamic environment selection of dataset. In Chuanmin, Hu et al.'s [44] works use slick region statistical parameters like mean, standard deviation, and minimum and maximum distance of the slick from the region. The results show an annual seapage rates in the northwestern Gulf of Mexico and also the slick time series variation recording. But the observed limitation is that it could be performed only on medium resolution of MODIS data under tropical and sub-tropical.

In 2010, Konstantinos Topouzelis et al. [45] have proposed an algorithm for feature selection and classification. In this algorithm, classification method uses the statistical decision tree rules and features considered are area, perimeter, complexity, shape, mean, power-to-mean ratio, texture, and contrast of oil spill region. The result shows that the obtained classification accuracy rate is 84.4% in detecting oil spill. This proposed method suffers from problem of classification of 70 trees and it performs only sequential selection operation.

In 2011, Biao Zhang et al. [46] have used an unsupervised classification approach for distinguishing oil slicks from sea surfaces. It uses very simple and effective mapping techniques. In this proposal, quad polarization SAR images of lower wind condition are mapped with normal condition by considering features such as slick entropy and slick region. But the observed limitation is the selected smaller dark regions on images could be processed only under moderate wind conditions on sea surface regions.

In 2012, Chanudhuri et al. [47] have proposed four step algorithms for automatic detection of ocean disturbance. This proposed approach first enhances SAR images to emphasize the dark regions. The second step then segments dark regions using iterative method. After this, the graph theory-based method uses to remove the unwanted false alarms. Finally, a link-based algorithm is applied for detecting the disturbance features using statistical approaches. In this algorithm, the used features enhance the image, segmentation of the image regions, hole filling, region removal of slick enhancement, segment link of the same regions, and spur removal of specified region. The result shows that synthetic test images with different noise levels and real images from a variety of satellites such as ERS-2, SEASAT, ENVISAT, RADARSAT for sea disturbance features. But this algorithm observed one limitation to compute the data models in a single way only. Bhogle et al. [48] have proposed algorithm for automatic detection of oil spills. This algorithm consists of three steps for protecting the water life and reduces environmental damages from oil spills. In this algorithm, the first step detects the dark spot on image regions. The next step is feature's extraction of each dark spot. The final step is classification of dark spot is oil spill or look-alike. After these steps, then the texture entropy is used for discrimination of oil spill area and Mahalanobis classifier is estimation of the oil spill identification. In this algorithm, authors used three features of oil spill such as the distance from coastline, mean of spill on specified region, and the co-variance of image region pixels. The reported result shows different algorithm for an automatic detection with accurate and efficient manner. And also computing the texture entropy features 94% of accuracy with help of Mahalanobis classifier. But this algorithm observed some limitations are texture entropy might complex process and less flexible using the Mahalanobis classifier is not effectively remove the speckle noise for data analysis. This proposed algorithm classifies the texture variations of spill detection on image regions.

In 2012, Michele Vespe et al. [49] have proposed approach for quality issues of satellite images related to marine applications. This algorithm is to assess the quality and indicators for vessel and oil spill detection. This algorithm used some features including global quality aspects of radiometric sensitivity, radiometric resolution, radiometric accuracy, region stability and error to detecting, spatial resolution, geolocation accuracy about the local-quality aspects with peak-to-side-lobe ratio, the integrated sidelobe ratio, ambiguity of slick detection, interference of artifacts, structured data, missing data, and SAR processing errors. The result is reported based on current issues of marine applications from oil spill detections. But the observed algorithm limitation to generalize the quality levels should be encountered based on the data representation.

In 2014, Alireza Taravat et al. [50] have proposed neural networks algorithm for fully automatic detection of darkspots. In this algorithm, two methods use non-adaptive identification of oil spill on image region. First method called Weibull Multiplicative Model (WMM) is applied for filtering in each sub-image. Second method called Pulse-Coupled Neural Networks (PCNN) is used for segment and removing the false targets. In this algorithm, authors used features as well-defined slick region identified, linear well-defined, massive well-defined, not well-defined, linear not well-defined, massive not well-defined, linear dark spot, and massive dark spot. The results tested 60 ENVISAT and ERS-2 images and reported 93.66% accuracy of overall dataset. But algorithm-observed limitation shows a fewer effective result.

In 2014, Yu Li et al. [51] have proposed a statistical quad-pol reconstruction method for SAR quad-polarization data. This method improves a compact polarimetric SAR data for reconstruction using statistical methods. This proposed method refines the scatter behavior of SAR signal relationship for iterative quad-pol reconstruction data. In this algorithm, authors used only one feature that is slick region iteration and the reconstruction of quad-pol data. The result shown is to improve compact polarimetric SAR data using the quad-pol reconstruction model accurately. But the algorithm observed one limitation is to perform only for low-accuracy quad-pol reconstruction model. This proposed algorithm has been used to improve the data performance and data accessing model for reconstruction using SAR polarimetric techniques.

In 2014, Salberg et al. [5] have proposed algorithm that uses hybrid polarimetric SAR technique for slick classification. This proposed algorithm is used only to improve hybrid-polarimetric SAR data and its feature extraction. In this algorithm, authors used five features that include oil spill detection on image dark regions, correlation co-efficiency, standard deviation, slick decomposition of region, conformity index of the slick with polarimetric SAR data, and coherence measurement of slick region separation. The result shows the features of hybrid polarimetric SAR data and then classifies each detected slick on the specified region. But algorithm observed one limitation that performs only limited testing images from the hybrid polarimetric SAR data. This proposed technique identifies slick feature and its measuring capabilities of hybrid polarimetric SAR data.

In 2014, Giacomo De Carolis et al. [52] have proposed an approach for measuring the slick thickness in medium-resolution imaging spectrometer instrument (MERIS) and moderate-resolution imaging spectro-radiometer (MODIS) images. This proposed approach estimates the thickness of oil slick from marine surface. In this proposed approach, authors used one feature for estimation of slick thickness from the detected region. The result shown is reported from the June to August 2006 under Lebanon oil spills occurred in MERIS and MODIS gathered data. But this approach not shows the perfect values of gathered data samples as a limitation.

In 2015, Collins et al. [53] have proposed approach as a compact polarimetric SAR technique for coherent dual-pol SAR images. This approach uses simulated airborne compact SAR data for characterizing oil-water mixing of deep-water horizon oil spill. On the other hand, this SAR is used to perform the great potential for maritime surveillance application on oil spill characterization. In this approach, author used single feature as reconstructing the errors for sea water. This result shows quad-pol data features and pseudo-quad data and their differences. But this approach is observed only for variations of different compact polarimetric SAR images.

In 2016, Saeed Chehresa et al. [54] have proposed an algorithm to detect the oil spills and lookalikes based on the selection of optimum features in a given SAR data set('s). This algorithm is evaluated based on classification of SAR image dark spots. This algorithm of 93.19% accurate is classified with optimum set of features from the dataset. Also, eight different evolutionary algorithms are considered to classify the desired feature subsets. Giacomo Capizzi et al. [55] have proposed an automate cluster-based system developed for oil spill detection in satellite remote sensing. This system interactively working with several characteristics about the availability and adaptability of the different classes of objects and SAR images are considered based on application areas. This proposed algorithm uses a back-propagation neural network algorithm to obtain the outcomes as identify objects such as ships with spills/sliks on SAR images.

In 2017, M. Konik and K. Bradtke [56] have proposed an object-oriented approach to detect oil spills on ENVISAT ASAR (Advanced Synthetic Aperture Radar) images. This proposed methodology improves the classification at the scale of entire water bodies, focusing on its repeatability. Also, this approach analysis enhances the optimized filters to multilevel hierarchical segmentation. This proposed system recorded 96.15% of accurately identified spills and 4% of dark spots extracted from the given dataset.

In 2018, Majidi Nezhad et al. [57] carried case study on sentinel 2 satellite images for oil spill detection analysis. In this analysis part, exploit areas of oil spills and oil loading ports are often used by the governance of tankers and ship. With ENVI tool, this analysis is carried out to discuss oil spill detection in Persian Gulf by using multi-sensor images data.

In 2019, Jiao et al. [58] proposed a deep learning approach to detect the oil spills in an unmanned aerial vehicle. This proposed approach having three main steps, first step used with CNN model to detect oil spills in image. Second step is used for filtering the detected results of first step with help of Otsu algorithm. Third step is used to detect detail manner in a region-based detection with oil spills. So, this proposed approach is effectively solved to reduce the cost of oil spill detection by 57.2% compared with the traditional inspection process.

In 2020, Yekeen et al. [59] proposed a novel deep learning algorithm to detect the oil spills. This process can be achieved by the similar visuals of oil slicks and look-alike, which affects the reliable SAR images for the marine oil spill detection. But, this can be used for minimal detection and discrimination of oil spills with help of traditional deep learning models with limited accuracy. Moreover, the proposed novel deep learning algorithm detects the oil spills using computer vision instance segmentation mask region based on CNN model to detect maximum oil spill on ocean regions.

In 2021, Syedi et al. [60] have proposed multiscale multidimensional residual kernel convolution neural network. This proposed method is used for overall accuracy of oil spill detection in Gulf of Mexico regions. Also, this study investigates a proposed model with OSD algorithms had shown better performance.

6. Classification using features of oil slick

In general, oil spill detection could be carried out either by manual process using trained personnel or by an automated system. As per literature review, existing systems uses either manual approach or semi-automated system. The numbers of SAR images to be analyzed are getting increased annually. If oil spill detection is carried out manually in a wide swath image with appropriate resolution, it is a very time-consuming process. Thus, it is required to design an automated system for improving results. In certain cases, as outlined in [31], it is observed that manual detection is preferred.

The manual oil spill detection approach at KSAT is described in [61]. About the wind speed Automatic Information System (AIS) ship tracking of sea and their outer surface information, location of oil rigs, pipeline, national territory borders, and coastlines are available to the operator. Possible oil spills are assigned confidence levels based on their contrast to the surroundings, wind speed, identification of possible sources, natural slicks or low-wind areas nearby, and edge and shape characteristics of the slick. Various algorithms for oil spill detection based on single-polarization SAR images are reviewed in [62–64]. Several of the papers describe a methodology consisting of identifying dark spots followed by computing features describing the shape, contrast, surroundings, and homogeneity of the spots. The main challenge is not to segment out dark areas in the SAR image, but to identify a set of good features that can be used to discriminate between oil slicks and look-alikes, and then to use the features in a reliable classifier.

Some of the algorithms for oil spill detection based on SAR polarization images were reviewed from a single direction. Discernibly, the huge uniqueness spill and sea

Sl. no.	Features
1.	Dark area (A)
2.	Dark area border (B)
3.	Ratio (A/B)
4.	Spill complexity
5.	Spill region shape
6.	Spill width
7.	Standard deviation of dark region
8.	Standard deviation of background
9.	Max dissimilarity between dark region and background
10.	Max dark area border gradient
11.	Mean dissimilarity between dark region and background
12.	Mean dark area border gradient
13.	Gradient standard deviation
14.	Local region contrast ratio
15.	Spill power-to-mean ratio
16.	Average of inside dark area
17.	Average of outside dark area
18.	Number of detected spots in the SAR image
19	Number of neighboring spots
20.	Gradient of dark area border

Table 5.
SAR image single-polarization features.

surface territories are checking of extensive regions for oil detection. The dissimilarity decreases about the generality of wind levels. But the main complicating thing is the low wind dark area with very high dissimilarity common in low-wind conditions. Most important features are, if possible, spills from the selected region can be seen on SAR image. Polarimetric SAR image features can be extended to a certain region and availability of SAR-limited data.

The existing system feature extraction for the oil spill detection is given in [63]. Most commonly used features are shown in **Table 5** for single-polarization SAR oil-spill detection and **Table 3** features are used for polarimetric SAR oil spill detection [64]. The most of polarimetric SAR features are listed in **Table 6**. These are not used for automatic algorithms. But some automatic methods are to be used for detection of oil spill with polarimetric SAR data. The main comparison between the accuracy with single SAR polarimetric and polarimetric SAR data is not yet processed for large datasets.

The existing system experimental results are tried to get the rank features according to their importance. The objective was to select and use only those features characterized by strong discriminative capacity. It appears that combinations of features with high discriminative capacity were not giving satisfying results as combinations of features having lower discrimination capabilities. For example, if we have 10 cases, three of which can be discriminated by feature X, the question is how many of the remaining seven cases can be discriminated by another feature, for example, feature

Sl. no.	Feature
1.	Entropy
2.	Mean scatters
3.	Standard deviation of CDP
4.	Coherent circular polarization
5.	Combined features
6.	Correlation coefficient

Table 6.
Polarimetric features used for oil spill detection.

Y. If features X and Y contribute only in discriminating the same three cases, then the combination is not good and another feature (e.g., feature Z) has to be used.

7. Current operational efforts in oil spill monitoring

The use of satellite imagery for regular oil spill monitoring worldwide is increasing. Pedersen et al. [65] describe the Canadian oil spill monitoring efforts. They currently acquire and analyze 5000 SAR images over Canadian waters annually. From 2006 to 2007, 360 images were analyzed, 12 anomalies were detected, and three of them were verified. An overview of long-term oil spill monitoring efforts in Europe is described. The Baltic Sea is an area where oil spill monitoring has been performed for many years already. Oil spill statistics are computed annually. According to the number of detected and verified oil spills correlates well with the shipping density. The European Maritime Safety Agency (EMSA) is responsible for providing vessel traffic monitoring services to the European Union Member States. EMSA manages maritime shipping systems: SafeSeaNet, CleanSeaNet, and LRIT. SafeSeaNet is a ship, vessel monitoring system based on AIS and LRIT track vessels outside the range of AIS coastal networks [66]. Thus, the CleanSeaNet offers near-real-time oil-spill detection from [67].

The EMSA CleanSeaNet satellite monitoring service provides annual statistics of potential oil slicks. In 2008, 3196 potential slicks were reported, while this was reduced to 2107 in 2009 and 1981 in 2010. In 2009, 751 of the reported slicks were checked by the national authorities and 194 verified (for verifying oil spills weekly) [68]. Although it is impossible to quantify the exact volume of oil involved in these spills, it is believed that deliberate discharges account for a progressively greater proportion of pollution than the accidental events. CleanSeaNet uses SAR images from ENVISAT, RADARSAT-1, and RADARSAT-2. The service is supposed to be extended [69], to include optical images (MODIS Aqua) in 2011. Operators assess the SAR images together with meteorological, oceanographic, and ancillary information to detect the presence of oil on the sea surface. Detection results are reported to the affected country less than 30 min after satellite image acquisition.

In case of a major accidental spill, European Union Member States and/or the European Commission will normally activate the International Charter for Space and Major Disasters [68]. EMSA can assist in analyzing satellite imagery to help coordinate

Figure 1.
This is an example of discharging oil from the ship.

Figure 2.
The Texas City accident on U.S. coast in March 22nd, 2014 © Houston Ship Channel.

and order all possible satellite imageries to continuously monitor the evolution of an accidental spill as well as to assist in applying forecasting computer models (**Figure 1**).

In the current operational scenario based on monitoring, the pollution control phenomena as aircraft and ship surveillance under the traditional techniques have many drawbacks: time delay, weather conditions, airborne surveillance, and SAR data evidence rules. These drawbacks are frequently reported by oil spills and their detection capabilities very low of the remote sensing monitoring. The Side-Looking Airborne Radar (SLAR) aircrafts observing is to detect pollution under different conditions about the wider areas of the coastal region surface [70]. However, the National Pollution Control Authorities Center for environmental protection detects the spoils from the oil spill damages: early warning, legal prosecution, and provision of pollution statistics (**Figure 2**).

8. Case studies

Outline of Oil Spill Detecting in Remote Sensing—The study of the major remote sensing gives major information about oil spills in marine surroundings or sea surface areas after accidents. Normally, the oil spills are caused by enormous accidents and their size of the oil area is covered from discharge region. After happening accidents, the huge regions can be covered by oil. The oil spill detection of accidents compared with smaller discharges will be different as the expected area of the oil spill will be different as the expected area of the oil spill will variation of the expected area of the oil spills in two different cases. The case of an accident is major assessment and the tracking is possible for oil trajectory models. gCaptain [63] shows a coast Guard photo from U.S. National Transporation Safety Board (NTSB) accident docket investigation on March 22nd, 2014 in the Houston Ship Channel (HUC).

The HUS is involving a cargo ship and a tank barge being measured by a Kirby tugboat. Its abound cargo ship MV Summer Wind collided with two-barge tow being led by the Kirby 27,706 tank barge and also pushed by the Kirby inland tow-boat Miss Susan at approximately 12:35 pm. In this case, most of the oil not immediately reaches the coastlines. The resulting collision is release of approximately 4000 barrels (168,000 gallons) of fuel oil into the waterway from a breach in the double hull of the Kirby 27,706 [71]. However, docket contains more than 4000 pages and it includes a summary of the oil incidents, transcripts from both vessels and witnesses aboard nearby vessels, and their factual report investigation. In 2002, many studies should be describing remote sensing of the Reputation accident; see the examples of [72]. The different satellites are such as ERS-2, ENVISAT, ASAR, RADARSAT-1, TerraSAR-X, and ALOS PALSAR data after oil spill incident on coastal. So, they associated the theoretical checking ratio from given frequency and its ratio estimated wind speed measurement. Normally, the sea surface conditions were not well appropriate to regularly detect surface oil out layer. But this could be covering a lack of sunglint the oil was difficult to show on visible band spectrum.

9. Conclusion

Oil slicks on the ocean surface are watched generally regularly. Contamination because of either mischances or think slick releases from boats speaks to a genuine danger on the marine environment. The effect of the spills relies on the sort and measure of oil, area, season, sea profundity, and meteorological and maritime conditions. Spaceborne remote detecting sensors can recognize spills before they cause broad harm. If there should be an occurrence of bigger mishaps, remote detecting symbolism gives simple reviews of the degree of the oil slicks. A mix of airborne and satellite-based remote detecting is at present utilized for operational oil slick checking around the world. Spaceborne SAR gives a diagram of substantial sea regions, and reconnaissance flying machine can be coordinated to check conceivable oil slick areas to confirm the spill and catch the polluter. Oil slick identification is most adequately performed on a substantial scale utilizing SAR pictures because of its every single climate capability [73] (given wind speeds in the reach 2–14 m/s) and great scope. The SAR scope relies on upon scope and is great in northern districts, where the climate conditions seriously restrain the functional utilization of optical sensors.

The most genuine downside with oil slick observing taking into account SAR pictures is that occasionally in low-wind conditions oil slicks cannot be dependably isolated from clones. In the course of the most recent years, a few new remote-detecting satellites with great oil slick location capacities have been propelled. Notwithstanding giving better scope as the quantity of operational satellite sensors expands, late research demonstrates that they additionally can give better separation between oil slicks and biogenic movies or carbon copies. Lately, various polarimetric measures valuable for oil slick discovery have been proposed. These incorporate both quad-pol elements such as polarimetric entropy and anisotropy, mean diffusing edge, polarimetric compass, congruity coefficient, and also the double pol components, for example, standard deviation of the co-energized stage distinction and the co-captivated relationship coefficient. By utilizing, the polarimetric SAR as accessible on RADARSAT-2 and TerraSAR-X, it appears that oil slicks can be recognized from biogenic movies. Optical sensors with great scope, as ENVISAT MERIS and Terra/Aqua MODIS, can give a decent supplement at geological areas where the likelihood of a without cloud scene with sunglint conditions is genuinely high.

Author details

Singanamalla Vijayakumar
Jain University, India

*Address all correspondence to: vkchinna5a8@gmail.com

References

[1] Frate FD, Petrocchi A, Lichtenegger J, Calabresi G. Neural networks for oil spill detection using ERS-SAR data. Geoscience and Remote Sensing. 2000;**38**(5):2282-2287

[2] Gade M, Redondo JM. Marine pollution in European coastal waters monitored by the ERS-2 SAR: A comprehensive statistical analysis. In: OCEANS'99 MTS/IEEE. Riding the Crest into the 21st Century. Vol. 3. 1999. pp. 1239-1243

[3] Walterscheid I, Klare J, Brenner AR, Ender JH, Loffeld O. Challenges of a bistatic spaceborne/ airborne SAR experiment. EUSAR. 2006:1-4

[4] Mccandless WWS, Jackson CR. Principles of synthetic aperture radar. SAR Mar. User's Man. 1978:1-23

[5] Franceschetti G, Lanari R. Synthetic Aperture Radar Processing. CRC Press; 1999

[6] Vijayakumar S, Santhi V. Different approaches for oil spill detection in SAR images – A review. International Journal of Oceans and Oceanography. 2015;**9**(2):221-228

[7] Solberg SHA, Dokken ST, Solberg R. Automatic detection of oil spills in ENVISAT, Radarsat and ERS SAR images. Geoscience and Remote Sensing Part-C. 2003;**4**:2747-2749

[8] Hovland AH, Johannessen AJ, Digranes G. Slick detection in SAR images. In: Proc. IEEE Symp. Geosci. Remote Sensing (IGARSS), Pasadena, CA. 1994. pp. 2038-2040

[9] Pedersen JP, Seljelv GL, Bauna T, Strøm DG, Follum AO, Andersen HJ, et al. Towards an operational oil spill detection service in the mediterranean? The Norwegian experience: A pre-operational early warning detection service using ERS SAR data. Spill Science & Technology Bulletin. 1996;**3**(1):41-46

[10] Pedersen JP, Seljelv GL, Strøm DG, Follum AO, Andersen HJ, Wahl T, et al. Oil spill detection by use of ERS SAR data. ERS Applications. 1996;**383**:181-185

[11] National Oceanic and Atmospheric Administration. Oil Spills: A Major Marine Ecosystem Threat. n.d. Available from: https://www.noaa.gov/explainers/oil-spills-major-marine-ecosystem-threat#:~:text=Depending%20on%20the%20circumstances%2C%20oil,them%20to%20the%20harsh%20elements [Accessed: August 12, 2022]

[12] Fingas M. The Basics of Oil Spill Cleanup. USA: CRC Press LLC; 2001

[13] Trivero P, Bimino W, Nirchio F. High resolution COSMO-SkyMed SAR images for oil spills automatic detection. In: Geoscience and Remote Sensing Symposium, 2007. IGARSS 2007. IEEE International; 2007. pp. 2-5

[14] Bringi VN, Hendry A. Technology of Polarization Diversity Radars for Meteorology. American Meteorological Society; 1990

[15] Reed M, Gundlach E, Kana T. A coastal zone oil spill model: development and sensitivity studies. Oil and Chemical Pollution. 1989;**5**(6):411-449

[16] Solberg AH. Remote sensing of ocean oil-spill pollution. Proceedings of the IEEE. 2012;**100**(10):2931-2945

[17] Moon WM, Staples G, Kim DJ, Park K. RADARSAT-2 and coastal

applications: Surface wind, waterline, and intertidal flat roughness. Proceedings of the IEEE. 2010;**98**(5):800-815

[18] Environmental Pollution Environment Pollution Effects: Oil Spill Pollution. 2009. Available from: http://www.environ mentalpollutioncenters.org/oil-spill/

[19] Emergency Management Emergency Current Information Management: Sensitivity of Freshwater Habitats. 2016. Available from: http://archive.epa.gov/emergencies/content/learning/web/html/freshwat.html

[20] Fingas M. Oil Spill Science and Technology. Gulf Professional Publishing; 2010

[21] Safe Environment Safe Environment: Oil Spill Effects on Marine Environmental System and Control Measures. 2008. Available from: https://saferenvironment.Wordpress.com/2008/09/17/oil-spill-adverse-effects-on-marine-environmental-bio-system-and-control-measures/

[22] Environmental Engineering. Environmental Engineering: Oil Spill Effects. 2008. Available from: http://www.environmentengineering.blogspot.in/2008/04/oil-spill-and-its-adverse-effects-on.html

[23] Emergency Management Emergency Management: Learning System of Oil Spill Habitats. 2016. Available from: http://archive.epa.gov/emergencies/content/learning/web/html/habitats.html

[24] Rice SD, Moles A, Taylor TL, Karien JF. Sensitivity of 39 Alaskan marine species to cook inlet crude oil and No. 2 fuel oil. International Oil Spill Conference. 1979;**1979**(1):549-554

[25] Wahl T, Skøelv A, Pedersen JP, Seljelv LG, Andersen JH, Follum OA, et al. Radar satellites: A new tool for pollution monitoring in coastal waters. Coastal Management. 1996;**24**(1):61-71

[26] Schumann G, Di B, G. The direct use of radar satellites for event-specific flood risk mapping. Remote Sensing Letters. 2010;**1**(2):75-84

[27] Adamo M, De Carolis G, De Pasquale V, Pasquariello G. Detection and tracking of oil slicks on sun-glittered visible and near infrared satellite imagery. International Journal of Remote Sensing. 2009;**30**(24):6403-6427

[28] Deguchi T, Kinugasa Y, Kurita K. Monitoring of land deformation using Terra SAR-X data around active fault in the metro Manila, the Philippines. InFIG Working Week; 2011

[29] Li ZH. Correction of Atmospheric Water Vapour Effects on Repeat-Pass SAR Interferometry Using GPS, MODIS and MERIS Data [PhD diss.]. University of London; 2005

[30] Science. Science & Philosophy: Octadecanoate to Ovenbirds. 2016. Available from: http://science.jrank.org/pages/4850/Oil-Spills.html

[31] Solberg AH, Storvik G, Solberg R, Volden E. Automatic detection of oil spills in ERS SAR images. Geoscience and Remote Sensing. 1999;**37**(4):1916-1924

[32] Marghany M. RADARSAT for oil spill trajectory model. Environmental Modelling & Software. 2004;**19**(5):473-483

[33] Galland F, Réfrégier P, Germain O. Synthetic aperture radar oil spill segmentation by stochastic complexity minimization. Geoscience and Remote Sensing Letters. 2004;**1**(4):295-299

[34] Keramitsoglou I, Cartalis C, Kiranoudis CT. Automatic identification

of oil spills on satellite images. Environmental Modelling & Software. 2006;**21**(5):640-652

[35] Girard-Ardhuin F, Mercier G, Collard F, Garello R. Operational oil-slick characterization by SAR imagery and synergistic data. Oceanic Engineering. 2005;**30**(3):487-495

[36] Mercier G, Girard-Ardhuin F. Partially supervised oil-slick detection by SAR imagery using kernel expansion. Geoscience and Remote Sensing. 2006;**44**(10):2839-2846

[37] Topouzelis K, Karathanassi V, Pavlakis P, Rokos D. Detection and discrimination between oil spills and look-alike phenomena through neural networks. ISPRS Journal of Photogrammetry and Remote Sensing. 2007;**62**(4):264-270

[38] Solberg AH, Brekke C, Husøy PO. Oil spill detection in Radarsat and Envisat SAR images. Geoscience and Remote Sensing. 2007;**45**(3):746-755

[39] Migliaccio M, Gambardella A, Tranfaglia M. SAR polarimetry to observe oil spills. Geoscience and Remote Sensing. 2007;**45**(2):506-511

[40] Chang L, Tang ZS, Chang SH, Chang YL. A region-based GLRT detection of oil spills in SAR images. Pattern Recognition Letters. 2008;**29**(14):1915-1923

[41] Brekke C, Weydahl DJ, Helleren Ø, Olsen R. Ship traffic monitoring using multi-polarisation satellite SAR images combined with AIS reports. In: Synthetic Aperture Radar (EUSAR), 2008 7th European Conference on VDE. 2008. pp. 1-4

[42] Ivanov AY, Zatyagalova VV. A GIS approach to mapping oil spills in a marine environment. International Journal of Remote Sensing. 2008;**29**(21):6297-6313

[43] Topouzelis KN. Oil spill detection by SAR images: Dark formation detection, feature extraction and classification algorithms. Sensors. 2008;**8**(10):6642-6659

[44] Hu C, Li X, Pichel WG, Muller-Karger FE. Detection of natural oil slicks in the NW Gulf of Mexico using MODIS imagery. Geophysical Research Letters. 2009;**1**:36-41

[45] Topouzelis K, Psyllos A. Oil spill feature selection and classification using decision tree forest on SAR image data. ISPRS Journal of Photogrammetry and Remote Sensing. 2012;**68**:135-143

[46] Zhang B, Perrie W, Li X, Pichel WG. Mapping Sea surface oil slicks using RADARSAT-2 quad-polarization SAR image. Geophysical Research Letters. 2011;**38**(10)

[47] Chaudhuri D, Samal A, Agrawal A, Mishra A, Gohri V, Agarwal RC. A statistical approach for automatic detection of ocean disturbance features from SAR images. Selected Topics in Applied Earth Observations and Remote Sensing. 2012;**5**(4):1231-1242

[48] Bhogle P, Patil SO. Oil spill detection in SAR images using texture entropy algorithm and mahalanobis classifier. International Journal of Engineering Science. 2012;**4**:4823-4826

[49] Vespe M, Greidanus H. SAR image quality assessment and indicators for vessel and oil spill detection. Geoscience and Remote Sensing. 2012;**50**(11):4726-4734

[50] Taravat A, Latini D, Del, and Frate, F. Fully automatic dark-spot detection from SAR imagery with the

combination of nonadaptive weibull multiplicative model and pulse-coupled neural networks. Geoscience and Remote Sensing. 2014;**52**(5):2427-2435

[51] Li Y, Zhang Y, Chen J, Zhang H. Improved compact polarimetric SAR quad-pol reconstruction algorithm for oil spill detection. Geoscience and Remote Sensing Letters. 2014;**11**(6):1139-1142

[52] De Carolis G, Adamo M, Pasquariello G. On the estimation of thickness of marine oil slicks from sun-glittered, near-infrared MERIS and MODIS imagery: The Lebanon oil spill case study. Geoscience and Remote Sensing. 2014;**52**(1):559-573

[53] Collins MJ, Denbina M, Minchew B, Jones CE, Holt B. On the use of simulated airborne compact Polarimetric SAR for characterizing oil–water mixing of the Deepwater horizon oil spill. Selected Topics in Applied Earth Observations and Remote Sensing. 2015;**8**(3):1062-1077

[54] Chehresa S, Amirkhani A, Rezairad GA, Mosavi MR. Optimum features selection for oil spill detection in SAR image. Journal of the Indian Society of Remote Sensing. 2016;**44**(5):775-787

[55] Capizzi G, Sciuto GL, Woźniak M, Damaševicius R. A clustering based system for automated oil spill detection by satellite remote sensing. In: International Conference on Artificial Intelligence and Soft Computing. Cham: Springer; 2016. pp. 613-623

[56] Konik M, Bradtke K. Object-oriented approach to oil spill detection using ENVISAT ASAR images. ISPRS Journal of Photogrammetry and Remote Sensing. 2016;**118**:37-52

[57] Kolokoussis P, Karathanassi V. Oil spill detection and mapping using sentinel 2 imagery. Journal of Marine Science and Engineering. 2018;**6**(1):4

[58] Jiao Z, Jia G, Cai Y. A new approach to oil spill detection that combines deep learning with unmanned aerial vehicles. Computers & Industrial Engineering. 2019;**135**:1300-1311

[59] Yekeen ST, Balogun A-L, Wan KB, Yusof. A novel deep learning instance segmentation model for automated marine oil spill detection. ISPRS Journal of Photogrammetry and Remote Sensing. 2020;**167**:190-200

[60] Seydi ST, Hasanlou M, Amani M, Huang W. Oil spill detection based on multiscale multidimensional residual CNN for optical remote sensing imagery. IEEE Journal of Selected Topics in Applied Earth Observations and Remote Sensing. 2021;**14**:10941-10952. DOI: 10.1109/JSTARS.2021.3123163

[61] Brekke C, Solberg AH. Oil spill detection by satellite remote sensing. Remote Sensing of Environment. 2005;**95**(1):1-3

[62] Brekke C, Solberg AH. Classifiers and confidence estimation for oil spill detection in ENVISAT ASAR images. Geoscience and Remote Sensing Letters. 2008;**5**(1):65-69

[63] gCaptain. gCaptain: Accidents on Oil Spills. 2015. Available from: http://gcaptain.com/ntsb-opens-accidentdocket-2014-houston-ship-channel-collision-oil-spill

[64] Itopf. Knowledge Resources: Oil Spill Fates. 2014. Available from: http://www.itopf.com/knowledge-resources/documents-guides/fate-of-oil-spills/

[65] Pedersen J, Seljelv L, Strm GD, Follum OA, Andersen JH, Wahl T, et al. Oil spill detection by use of ERS SAR

data; from R&D towards pre-operational early warning detection service. In: Proceedings of the Second ERS Applications Workshop. London; 1996

[66] Tenecoport. Tenecoport 3rd Part: 8th Forum Monitoring and Control Systems. 2014. Available from: http://www.tenecoport.eu/images/3rd_ecoport_8_forum_MonitoringControl-ships.pdf

[67] Atkins JP, Burdon D, Elliott M, Gregory AJ. Management of the marine environment: Integrating ecosystem services and societal benefits with the DPSIR framework in a systems approach. Marine Pollution Bulletin. 2011;**62**(2):215-226

[68] Grüner K, Reuter R, Smid H. A new sensor system for airborne measurements of maritime pollution and of hydrographic parameters. GeoJournal. 1991;**24**(1):103-117

[69] Earth Workshop. Earth Workshop Seasar: Operational Uses of SAR Satellites. 2010. Available from: http://earth.eo.esa.int/workshops/seasar2010/5_chintoa_uta.pdf

[70] NTSBInvestigation. NTSB Accident Reports: Marine Investigation and Accident Reports. 2014. Available from: http://www.ntsb.gov/investigations/AccidentReports/Reports/MAR1501.pdf

[71] Costanza R. Ecological Economics: The Science and Management of Sustainability. Columbia University Press; 1992

[72] Hühnerfuss H, Alpers W, Jones WL. Measurements at 13.9 GHz of the radar backscattering cross section of the North Sea covered with an artificial surface film. Radio Science. 1978;**13**(6):979-983

[73] Parameswaran V, Burlina P, Chellappa R. Performance analysis and learning approaches for vehicle detection and counting in aerial images. Proceedings of IEEE International Conference on Acoustics, Speech, and Signal. 1997;**4**:2753-2756